W9-CGQ-411

DENSE CHLORINATED SOLVENTS

IN POROUS AND FRACTURED MEDIA

DENSE CHLORINATED SOLVENTS

IN POROUS AND FRACTURED MEDIA

MODEL EXPERIMENTS

BY FRIEDRICH SCHWILLE

WITH ASSISTANCE FROM
WOLFMAR BERTSCH, RENATE LINKE
WALTER REIF, SIGMUND ZAUTER

TRANSLATED BY JAMES F. PANKOW

ENGLISH LANGUAGE EDITION

Library of Congress Cataloging-in-Publication Data

Schwille, Friedrich.
Dense chlorinated solvents in porous and fractured media.

Translation of: Leichtflüchtige Chlorkohlenwasserstoffe in porösen und klüftigen Medien.
Includes index.
1. Water, Underground—Pollution—Experiments.
2. Organochlorine compounds—Environmental aspects—Experiments. 3. Solvents—Environmental aspects—Experiments. 4. Porous materials—Experiments.
I. Bertsch, Wolfmar. II. Title.
TD426.S3813 1988 628.1'6836 87-29679
ISBN 0-87371-121-1

LEWIS PUBLISHERS, INC.
121 South Main Street, Chelsea, Michigan 48118

PRINTED IN THE UNITED STATES OF AMERICA

This translation is dedicated to my parents,
Bernard John Pankow and
Irma Lillian (Cordts) Pankow.

Contents

Biographical Sketches ix

Foreword xiii

Translator's Preface xvii

Preface to the English Edition xix

Preface to the German Edition xxi

Acknowledgments xxiii

List of Figures xxv

List of Tables xxix

Abbreviations, Notation, Symbols xxx

1 Theoretical Considerations 1

2 Spreading Models 5

3 Spreading as a Fluid Phase in Porous Media 9
- 3.1 Preliminary Experiments 9
- 3.2 Trough Experiments 12
- 3.3 Column Experiments with Low Permeability Saturated Media 19
- 3.4 Glass Frit Experiments 24
- 3.5 Lysimeter Experiment 28

4 Retention Capacities of Porous Media 39

5 Microscopic Examinations 47

5.1 Methods 47
5.2 Examples 51
5.3 Interpretation 57

6 Spreading as a Fluid Phase in a Fractured Media 61
6.1 Methods 61
6.2 Examples 66

Color Plates I through XXIV 73
Dense Chlorinated Solvents in Porous and Fractured Media
Model Experiments: 48 Full-Color Photographs

7 Migration in Aqueous Solution 99
7.1 Dye Experiments on Density-Affected Flow in Porous Media 99
7.2 Dye Experiments in Fractured Media 103
7.3 Solubilization: Removal of CHCs at Residual Saturation in Porous Media 103
7.4 Solubilization: Removal of CHCs in Pools 108
7.5 Adsorption-Desorption in Porous Media 112

8 Spreading as a Gas Phase 119

9 Conclusions 125

Appendix I: Translator's Appendix: Physical and Chemical Properties of Dense Solvent Compounds 129

Bibliography 133

Index 145

Dr. Friedrich Schwille received his doctorate in 1950 in physical science from the Darmstadt University of Technology in the Federal Republic of Germany. He was then employed as a hydrogeologist at the Geological Survey of Rheinland-Palatinate where he worked largely in the fields of groundwater exploration and groundwater protection. Since 1960, Dr. Schwille was chief groundwater hydrologist at the Federal Institute of Hydrology in Koblenz. He was responsible for dealing with subsurface water issues and problems that arose during the planning, improvement, and construction of waterway systems by the Federal Ministry of Transport. He also directed many investigations for the Ministry of Interior on the migration of contaminants in groundwater systems. Chemical materials of special interest in this context were fluids that are immiscible with water, including petroleum-related products and dense halogenated solvents. Though now retired in Koblenz, Dr. Schwille maintains an active and continuing interest in the issues related to the contamination of groundwater, and travels frequently within Europe as well as to North America to discuss his research results with the growing number of persons who are interested in his pioneering work.

Dr. James F. Pankow is a professor, and chairman of the Department of Environmental Science and Engineering, and also the director of the Water Research Laboratory at the Oregon Graduate Center in Beaverton, Oregon. He received his BA in chemistry in 1973 from the State University of New York at Binghamton. His work at SUNY/Binghamton included research in ion exchange conducted under the direction of Dr. Gilbert E. Janauer. In 1979 he obtained his PhD in Environmental Engineering Science from the California Institute of Technology, working under the direction of Dr. James J. Morgan. His research group at OGC is involved in the study of the physical and chemical processes affecting the behavior of organic chemicals in the environment. These efforts encompass a wide range of both field and laboratory experimentation activities. An important component of this work is Dr. Pankow's research concerning the development and application of sensitive analytical methods for the determination of organic contaminants in the air and water environments. Dr. Pankow serves as a consultant to numerous federal agencies including the Department of Energy, the National Science Foundation, the U.S. Geological Survey, and the U.S. Environmental Protection Agency.

Dr. John A. Cherry received a bachelor's degree in geological engineering from the University of Saskatchewan in 1962, a master's degree in geological engineering from the University of California in 1964, and a PhD in geology from the University of Illinois in 1966. From 1967 to 1971 Dr. Cherry was an assistant professor and then associate professor in the Department of Earth Sciences at the University of Manitoba. His involvement in field studies of contaminant behavior in groundwater and investigations of subsurface waste disposal systems began in 1968. Since 1971 he has been a professor at the University of Waterloo and is now director of Waterloo's Institute for Groundwater Research. He is a coauthor (with R. A. Freeze) of the textbook *Groundwater*. He is the 1985 recipient of the Meinzer Award of the Geological Society of America and the 1985 recipient of the Horton Award of the American Geophysical Union for advances in the knowledge of the behavior of contaminants in groundwater. In 1986 he was selected as the first Henry Darcy Distinguished Lecturer for the National Water Well Association. For the past 17 years his research interests have focused on field studies of contaminant migration processes. He has been active as a consultant to various government agencies and various corporations in the United States and Canada.

Foreword

The contamination of groundwater systems by organic chemicals is an environmental issue that has been thrust upon developed nations very rapidly during recent years, seemingly without warning. Much of this contamination has taken place during the period of industrial development that followed World War II. It was born of a perception that subsurface systems possess unlimited capacities to accept and degrade organic chemicals introduced either accidentally or intentionally by disposal operations. This perception is now recognized as having been unfounded, as were the similar, early views of the waste tolerance capacities of surface waters, the oceans, and the atmosphere.

Dense halogenated solvents are now recognized as the cause of a large portion of the contemporary groundwater problem. These compounds are widely used in machine and electronics manufacturing, automotive and engine repair, airplane servicing, dry cleaning, asphalt operations, musical instrument manufacturing, dye manufacturing, and many other industrial operations. In addition, these solvents have been used in household products such as paint thinners and septic tank degreasers. The annual production of halogenated solvents is immense. In 1979, for example, the aggregate U.S. production of methylene chloride, 1,1,1-trichloroethane, trichloroethylene, and perchloroethylene was more than a million metric tons.

Considering the extensive uses that halogenated solvents have in industrialized society, it is not surprising that hundreds of Superfund sites in the United States are characterized by groundwater that is contaminated with these chemicals. While facilities that have handled large amounts of solvents have certainly been responsible for some of the current large-scale contamination problems, even a comparatively small leak of liquid solvent can cause very extensive contamination. For example, a plume that is 1000 m long, 100 m wide, and 20 m deep with an average concentration of 100 ppb (i.e., 20 times the U.S. EPA recommended limit for most solvents in drinking water) contains only 80 kg of contaminant, or the equivalent of 56 liters of a solvent like trichloroethylene. (Zero sorption to the aquifer has been assumed.) Obviously, then, a single 206-liter (55-gallon) drum of solvent can cause a very large problem.

The fact that groundwater contamination by halogenated solvents has come as a surprise is regrettable; much of the problem could have been avoided. Indeed, we realize in retrospect that information has long been available that pointed to the potential that these chemicals have for causing rapid and extensive contamination. In particular, we have known for many years that these compounds are in general: (1) more dense and less viscous than water; (2) not nearly as biodegradable as other organic compounds; (3) quite soluble relative to the low levels which require regulatory action; (4) largely nonsorbing and therefore quite mobile in groundwater systems; and (5) rather volatile.

With respect to their densities, it could have been more fully appreciated that, when spilled in adequate volumes, these compounds are capable of penetrating the capillary fringe and sinking deep into an aquifer system. Similarly, it was to be expected that the low viscosities of many of these compounds would only serve to facilitate both their initial percolation through the unsaturated zone as well as their subsequent entry into the saturated zone.

As far as their biodegradabilities are concerned, it is somewhat incongruous that the problems these chemicals are capable of causing were not anticipated, when, on the other hand, it has been surprising to learn in recent years that many of these compounds are capable of being biodegraded at all. This contrast becomes all the sharper when considered in light of the fact that soil scientists were aware many years ago that chemicals of this nature do not tend to partition significantly either to soil organic carbon or to soil mineral surfaces. As a result, usually only dispersion and biodegradation are left as mechanisms for the attenuation of concentrations which are, moreover, not limited to tolerable levels by the solubility limits of the halogenated solvents themselves.

With regard to the significant volatilities of these chemicals, it may be noted that it is this property that has led to the greatest degree to the lack of appreciation of their potential for causing groundwater contamination. Solvents spilled on, or disposed of near the ground surface were thought to be capable of volatilizing rapidly and completely to the atmosphere. Such volatilization was surely important in reducing the severity of many cases of solvent contamination. However, it must be remembered that in all such cases the unsaturated zone was initially free of those chemicals, and so the driving force for volatilization and diffusion *into* the unsaturated zone should have been expected to be large. Moreover, the presence of liquid water in the unsaturated zone into which the solvents could partition, and the subsequent infiltration of fresh water into a contaminated unsaturated zone could only have been expected to facilitate the spreading and increase the magnitude of the subsurface contamination.

Appreciable recognition of the halogenated solvent problem in North America did not come until the early 1980s. The lateness of this recognition was due to the fact that groundwater monitoring for volatile organic compounds and nearly all other organic compounds did not become common until that time. Regulatory agencies concerned with groundwater management had their attention fixed primarily on radioactivity, heavy metals, detergents and gross indicators of organic contamination such as COD, BOD, and TOC. Also, as already discussed, the groundwater research community had generally overlooked the potential magnitude of this problem.

Dr. Friedrich Schwille was one of the first scientists to recognize how the various properties of the dense halogenated solvents are capable of leading to severe groundwater contamination. Chief groundwater hydrologist at the Federal Institute of Hydrology in Koblenz, West Germany from

1960 to 1985, Dr. Schwille has been aggressively pursuing pioneering research in this topic area since 1975, i.e., well before the significance of the solvent problem was recognized in North America. Though now retired in Koblenz, Dr. Schwille maintains an active and continuing interest in the study of the impact of halogenated solvents on groundwater systems.

Seeking to develop both a qualitative and a quantitative understanding of the interplay of chemical and physical properties affecting the patterns and rates of contamination, Dr. Schwille has obtained much insight concerning these issues from an extensive series of physical model experiments conducted in Koblenz. The results of these experiments are extremely helpful in developing the type of intuitive understanding of halogenated solvent behavior that has heretofore often been lacking outside of Dr. Schwille's laboratory. This book contains a synopsis of those results.

The widespread recognition that Dr. Schwille is now receiving for his pioneering work is overdue. It is therefore our considerable pleasure to be able to present this book for use by the English-speaking community of scientists and engineers concerned with groundwater contamination by dense halogenated organic solvents.

James F. Pankow
Beaverton, Oregon

John A. Cherry
Waterloo, Ontario

Translator's Preface

Wherever possible, it has been the goal of this translation to maintain the same sentence and thought progression that exists in the German edition. In some instances, however, a minor amount of rearrangement served to improve the presentation of the ideas in English. While the translation of most of the specific terms was straightforward, it should be pointed out that the term *Rückhaltevermögen* has been translated to *retention capacity*. Dr. Schwille has used Rückhaltevermögen in two senses: (1) in a general sense, as the *overall capacity* of a given system or stratum to retain chlorinated solvents; and (2) in a specific sense, for the *residual saturation* (Residualsättigung) CHC content of a porous medium. For example, when an impermeable layer upon which a spilled CHC will collect is present within a given geological system, the retention capacity of that system will be greater than the residual saturation of the overlying porous medium. When no such barriers to CHC infiltration exist, then the retention capacity will equal the residual saturation. This duality of meaning for retention capacity has been retained in the translation, and each specific usage may be drawn from its context.

The translation of certain other aspects of the text was facilitated by helpful comments from Dr. Schwille. I thank Dr. Larry D. Wells and Dr. Gilbert E. Janauer for early instruction in the fine points of the German language. The assistance of Elizabeth Barofsky and Michael R. Anderson in eliminating errors is also gratefully acknowledged. I thank the German Federal Institute of Hydrology/Koblenz for authorizing this translation as well as for providing the original figures for use in this document.

This translation was funded in part from financial support provided by the IBM Corporation for groundwater research.

James F. Pankow
Beaverton, Oregon

Preface to the English Edition

This document provides a summary of the experimental research carried out at the German Federal Institute of Hydrology in Koblenz on the topic of how halogenated solvents behave in the subsurface environment. It has been a pleasant surprise for me to learn that this document has generated a substantial amount of interest in Canada and in the United States. It is in this context that Dr. John A. Cherry of the University of Waterloo suggested that this document should be translated into English in order to make it more available to English-speaking researchers.

Dr. James F. Pankow of the Oregon Graduate Center was willing to undertake the substantial task of carrying out the translation. It is with the greatest level of attention to detail that he has carried out this work. The English document that has resulted from his work is not a word-for-word translation, but rather a very thoughtful translation that preserves the intended meanings found in the original document. Indeed, Dr. Pankow has clarified substantially certain portions of the original text that were somewhat unclear in the German edition.

I hope that this translation will encourage researchers at other institutions to conduct further work in this field; some of the initial experiments that we carried out could well be expanded and carried further. It is with much pleasure that I thank Dr. John A. Cherry and Dr. James F. Pankow for their interest in our early work in this field.

F. Schwille
Koblenz

Preface to the German Edition

We began our model experiments concerning the behavior of volatile chlorinated hydrocarbons (CHCs) in the subsurface in 1975. At that time, we desired to conduct experiments that related to groundwater contamination on a massive scale. In particular, we were interested in spill events such as those that can occur in the storage and transport of these chemicals. The intent was to start to accumulate experience with CHCs. It would then be possible for the water studies field to avoid the situation that occurred with petroleum-related contaminants in the 1950s and 1960s, namely being confronted with "new" toxic materials with unfamiliar and difficult to explain fluid mechanical behaviors. As the experience with petroleum showed, it requires a considerable length of time for the water science establishment and water studies personnel to become familiar with the behavior of liquids that are completely different from water.

Within five years, during which we conducted experiments more or less along the way, we succeeded at least in experimentally validating predictions made on the basis of our experience with petroleum products concerning the basic principles governing the spreading of CHCs in the subsurface. Then, in the late 1970s, the research programs dedicated to the study of how trihalomethanes are formed by the chlorination of groundwater containing humic substances discovered that solvent-type CHCs were present in many groundwater systems. This came as a surprise to many. The water industry then had to act and initiate the needed directed research. In the summer of 1981, the Ministry for Nutrition, Agriculture, Environment, and Forestry at Baden-Württemberg initiated an appropriate research program immediately. The German Federal Institute of Hydrology was invited to take part in this project. This provided us with the opportunity to broaden the basic knowledge that we had gathered up to that point in time.

I could give a large number of lectures in many sessions on our experimental results to specialists in Germany as well as the surrounding countries. To my knowledge, similar experiments have not been carried out at other institutions. Therefore, this research field has, in a very large manner, adopted the perception concerning the spreading of CHCs that we have set forth. The fact that no serious objections have been raised against our interpretations is probably due to the fact that we occasionally documented the course of our experiments with films.

The impressive pictures have often been very convincing for others. In addition, and this was to be sure *not intended*, they have also partly brought some to the assumption that the problem of CHCs is now solved from the fluid dynamics point of view. However, it is my perception that while the basic principles of the spreading have been clarified, many questions of detail remain unanswered.

The publication that lies ahead attempts at this point, at the risk of repeating ourselves, to collect all of our previous results and to interpret them in the light of our newer results. At the same time, I am taking the opportunity to answer in a concrete manner many of the questions that have been posed. In addition, I will clarify misunderstandings that may tend to arise. Only when one knows how the model experiments have been designed can he transfer the results to other situations in a meaningful manner.

I would not like to neglect at this time to heartily thank the members of the "Water and Wastewater" Committee of the Chemical Industry Association (registered); their experience in dealing with chemicals was very valuable. In addition, I heartily thank my collaborators on the Baden-Württemberg research project for a continual flow of information, technical advice, and critical direction.

F. Schwille
Koblenz

Acknowledgments

The completion of the model experiments required a number of coworkers. They completed their tasks with patience, prudence, and conscientiousness. We thank Mrs. Käthe Bromba and Mrs. Gisela Diesler for the preparation and execution of the experiments. Mr. Klaus Güls and Mr. Volker Schwaniger are thanked for constructing and maintaining the model equipment. We also thank Mrs. Christa Tegen for the drafting of the figures and Mrs. Margit Breilung for the preparation of the manuscript.

List of Figures

1. Size distribution curves for model sands. 10

2. Sinking of TCE in sands of different permeabilities. 11

3. Schematic diagram of large model trough. 12

4. Spreading of the dyed water solution in the two-layered groundwater aquifer in large model trough. 13

5. Spreading of PER in large trough (Experiment I, intermediate stages). 14

6. Spreading of PER in large trough (Experiment I, final stage). 15

7. Spreading of PER in large trough (Experiment II, final stage). 15

8. CHC recovery by means of a well in a two-pump system. 17

9. Glass trough; groundwater with zero gradient. 18

10. Column experiment; saturated porous medium with low permeability. 20

11. Results obtained in Figure VIa experiment. 22

12. Diagrammatic sketch of a glass frit column with water level adjustment vessel. 25

13. Relationship between the pressure required for the beginning of breakthrough of PER through water-saturated glass frits and the frit pore size. ... 27

14. Lysimeter experiment; size distribution data. 29

15. Lysimeter experiment; development of the infiltration front for simulated fluorescein-dyed "precipitation." ... 30

16. Lysimeter experiment; curve for addition of PER, and curves for the outflow of PER and water. ... 31

17. Lysimeter experiment; development of the infiltration front for PER. ... 32

18. Lysimeter experiment; water ("precipitation") application and PER outflow. ... 34

19. Lysimeter experiment; PER distribution in selected cross sections. ... 35

20. Lysimeter experiment; retention capacity as a function of time. ... 36

21. Glass columns for residual saturation determinations. ... 40

22. The outflow of water from a water-saturated column as a function of dewatering time. ... 42

23. The progress of the PER phase front as a function of time. ... 43

24. Determination of the residual saturation of PER as a function of the drainage time. ... 43

25. The retention capacity as a function of the draining time. 44

26. Frame cell for examinations with a macroscope. 48

27. Fluid distributions in the unsaturated zone (schematic). 50

28. Displacement of PER in the unsaturated zone by water (schematic). 52

29. Sinking of PER in a saturated medium. 58

30. The relative permeability curves for water and a typical CHC in a porous medium as a function of the pore space saturation. 59

31. Fracture model experiment with PER during infiltration. 63

32. Fracture model experiment with PER during infiltration. 64

33. Fracture model experiment with PER during infiltration. 65

34. Schematic diagram of the large model trough as it was used in the dispersion experiments. 101

35. Results of the dispersion experiment with a 0.35-g/L fluorescein solution. 102

36. Schematic diagram of the simple solubilization experiment. 105

37. Results of the solubilization experiment. 105

38. Standard apparatus for the study of solubilization. 106

39. Schematic diagram of the flat trough for the determination of the removal of CHC from pools. 109

40. Adsorption-desorption experiments with PER, TCE, and DCM in a low-clay content quartz sand. 114

41. Experimental apparatus for the determination of the sinking of gaseous CHC in a porous medium. 120

42. Time-dependent gas-phase chloroform concentration in the lower glass reducer of the apparatus presented in Figure 41. 121

List of Tables

1. Pore sizes of glass filters used in glass frit experiments 25

2. Lower bound estimates for residual saturation 45

3. Average concentration of solubilized PER with three experimental sands 107

4. Results of the solubilization experiments in the flat trough 110

5. Experimental data 113

6. Breakthrough (BT) volumes, volumes to achieve $C/C_0 = 1$, and volumes to achieve $C/C_0 = 0$ 115

7. K_d values obtained from breakthrough experiments 117

8. BT values for CHC and NaCl from breakthrough experiments 117

9. True relative vapor densities at 20°C 122

Abbreviations, Notation, Symbols

C	concentration (mg/L)
c	fracture width (mm) (denoted as B in Figures XXIa–XXIIb)
CHC(s)	volatile chlorinated hydrocarbon(s)
DCM	dichloromethane (methylene chloride)
H	thickness of the model aquifer (cm)
h_c	height of the capillary fringe (cm)
K	hydraulic conductivity of a porous medium for water (m/sec)
n	porosity
PER	tetrachloroethylene (perchloroethylene)
T	experimental, i.e., room, temperature (°C)
1,1,1-TCA	1,1,1-trichloroethane
TCE	trichloroethylene
V_p	pore volume of a porous medium (L)
▽ ----	visible upper boundary of the capillary fringe, i.e., the boundary between the unsaturated and the saturated zones
▼ ——	groundwater table
ν	kinematic viscosity (mm^2/sec)
ρ	density (g/cm^3)

1

Theoretical Considerations

When chlorinated hydrocarbons (CHCs) were discovered in many groundwater systems at the end of the 1970s, the literature was deficient concerning their behavior. Indeed, the fact that CHCs were found in groundwater at all appeared rather contradictory. For example, on the one hand, one knew that CHCs are not miscible with water. On the other hand, investigations of spill situations seldom indicated the presence of the pure phase. Rather, only the dissolved form was mainly found. (Small amounts of dissolved CHCs were, by the way, not yet recognized to pose a real threat to groundwater: the prevailing opinion was that CHCs would not be capable, because of their volatilities, of penetrating deep into the subsurface.) How then do extensive plumes of groundwater contamination evolve from CHC spills that are originally small in scale?

Before we began our model experiments, we tried to predict the behavior of the CHCs in the subsurface on the basis of their well-known, available, physicochemical data. The results were that these low-molecular weight materials, used predominately as solvents, form a rather uniform group. Moreover, they have much in common with fuels used with carburetors: they are practically insoluble with water but nevertheless possess water solubilities that are of the same order of magnitude as those of carburetor fuels; and, with only one exception, they evaporate very quickly. In addition, the surface tensions of CHCs between water and air lie in the same ranges as apply to carburetor fuels. There is only one significant difference: the CHCs are substantially heavier than water.

Would it be possible, therefore, for CHCs to sink through pores or fractures all the way to the bottom, confining layer of an aquifer? This question was by no means as easy to answer as many today believe. One can, for example, carry water (up to a given thickness) in an unwetted sieve. One can take a sewing needle and, with a little care, lay it on a water surface and have it float. One must therefore come to the conclusion that a CHC phase will usually be prevented (by the hypothetical surface tension skin that surrounds it) from entering at least the small pores of a water saturated medium, and that it will not be able to penetrate those pores unless it is under substantial pressure. Therefore, the above question was not to be answered without experimentation. It was revealed, nevertheless, that one can apply the principle of flow for two to three mutually immiscible phases originally developed for petroleum. That means that a CHC will spread primarily due to the influence of gravity until the point is reached at which the fluid no longer holds together as a single continuous phase, but rather lies in isolated residual globules, i.e., in the so-called condition of residual saturation. At that point, the CHC has become largely immobile under the usual subsurface pressure conditions and can migrate further only: 1) in water according to its solubility; or 2) in the gas phase of the unsaturated zone.

We were solidly convinced of this principle when we developed the first conceptual diagrams of the spreading of CHCs. We then improved them again and again with the results obtained from our progress until finally the depictions given in Figures IIIa–IVb came into being. (Those readers who are familiar with the literature will immediately recognize the numerous similarities between these figures and the plates we developed to depict the subsurface spreading of petroleum.) It goes without saying that we repeatedly tested the extent to which the results of the model experiments corresponded and agreed with findings obtained in the investigation of accidental spills. (The systematic spill reports of the Geological Survey of Baden-Württemberg were of special value to us in this regard.)

Before going into the model experiments, I will—in some ways partially anticipating the experimental results that have been obtained—briefly describe the spreading process by means of Figures Ia and IIIa–IVb. This will be done to better illustrate the purposes that our experiments served, as well as to point out the questions that have not yet been answered satisfactorily.

2

Spreading Models

A fluid-mechanically correct model for the *active spreading* and the *passive transport* of chlorinated hydrocarbons (CHCs), two terms that we will refer to collectively as *migration processes*, is essential if specialists from different disciplines want to both understand and make themselves understood. Even in conversations among specialists, careless or even false notation and concepts are unwittingly used again and again. For example, when one person uses the term *solution* to refer to a specific solvent CHC in its fluid phase form, another understands him to mean only a water solution of that CHC. A CHC present as a component in waste oil would also be a solution. Similarly, many times when a CHC is held in residual saturation, this condition will be referred to simply as an *adsorption* phenomenon; others will use the term adsorption to refer to chemical sorption to aquifer materials. Finally, according to preference, the term *saturation* is often used to refer to saturation of the medium with water, or to residual saturation with CHCs, or to both.

In place of the terms *seepage zone* and *aquifer*, we will often use *unsaturated zone* and *saturated zone*, respectively. The last two terms are being adopted increasingly by the groundwater field. When used alone, the term *saturation* will refer to water saturation and not to saturation with a CHC. In addition, the terms *dry* and *wet* will be used exclusively to refer to water.

Figure Ia

We have chosen specific colors to denote a CHC in different states (i.e., in the fluid phase, in solution in water, and in the gas phase). CHC in the fluid phase is represented with red. The lighter hues of red are for CHC residual saturation in the unsaturated (water) and saturated (water) zones, while the darker hues are for levels of CHC that are higher than CHC residual saturation. CHC dissolved in water is denoted with yellow and orange: yellow for the unsaturated zone and orange for the saturated zone. CHC in the gas phase is denoted with green: light green is for the lower concentrations in the outer reaches of the contamination, and dark green is for the higher concentrations at the core of the contamination. (In the model experiments, we used CHCs dyed with the organic compound oil red to distinguish the CHC phase. To distinguish a CHC dissolved in water, we used the dye fluorescein. This latter compound is bright green at low concentrations.) The color gray is for porous media; brown is for the impermeable bottom of an aquifer; blue is for uncontaminated water in the unsaturated and saturated zones.

Figure IIIa

If the unsaturated zone is fairly permeable to air, the majority of the CHC will evaporate there relatively quickly. A gas mound will therefore build itself around the body of the mass of CHC phase, and the concentration in that mound will decrease with increasing distance. The mass of the CHC phase will ultimately vaporize completely and thereby move into the gas mound: it will form a gas zone which holds together. As a result of their relatively high densities, the CHC gas will tend to sink. It will spread out laterally over strata which are not very permeable. The farthest the gas will sink is to the capillary fringe, over which it will then also spread out laterally. The scenario in Figure IIIa corresponds to the type of underground contamination that can occur in flat areas that are covered with roofs, or in very permeable soils during extended periods of little precipitation.

Figure IIIb

When the amount of CHC infiltrates is less than the capacity of the unsaturated zone, the fluid will spread itself out under the influence of gravity until it finally reaches the state of residual saturation. Under the conditions which are typical of subsurface systems, it will then be immobile. When the CHC collects on top of layers which are less permeable, it will accumulate to levels which are higher than the residual saturation. This will especially be the case when the less permeable layers are saturated with water. Therefore, in those cases when the CHC percolates relatively rapidly through permeable strata, the less permeable, fine-grained strata will lead to lateral spreading, particularly when those less permeable strata contain substantial water. The water which subsequently infiltrates through the body of the CHC mass will then transport dissolved CHC to the water table, or more precisely stated, to the capillary fringe. Since the CHC solution will, in general, be only slightly more dense than the uncontaminated groundwater, the dissolved CHC will then be carried along in a predominantly horizontal direction by the groundwater flow. A general term to describe the resulting contamination would be the *solution zone*. In the saturated zone, the term *solution plume* can also be used because a trail of contamination which extends away from the source is analogous to a smoke plume. (In addition to the generation of CHC contamination as depicted in Figure IIIb, one can also envision generation of contamination from *CHC-containing wastes* spilled either at the surface or below ground level.)

Figure IVa

A volatile CHC will form a vapor cloud (see Figure IIIa) that will surround the body of the CHC mass with a gas mound. Since the gas mound will usually extend significantly beyond the center of the CHC mass, the solution zone will also be significantly more extensive in both the unsaturated and saturated zones. In the unsaturated zone, the contaminated gas phase zone is also the contaminated solution zone.

Figure IVb

If the percolated CHC mass exceeds the retention capacity of the unsaturated zone, then, with sufficient pressure, the excess mass of CHC will force itself into the saturated zone. This will proceed until the residual saturation is reached there as well. As in the unsaturated zone, intervening impermeable strata in the saturated zone will obstruct the downward migration and promote the lateral spreading of a CHC. This process is even capable of stopping the sinking of the CHC. The effects of these impermeable strata are accentuated in the saturated zone since the presence of pore water greatly decreases a medium's permeability to a CHC. If the retention capacity of the saturated zone is also exceeded, the CHC will spread itself over the bottom of the aquifer in the form of low-lying mounds shaped like watch glasses. The CHC will also tend to collect in basins and depressions in the bottom of the aquifer, forming pools and puddles.

The groundwater that flushes through the CHC mass and flows over the pools and puddles on the bottom of the aquifer will transport solubilized CHC further in the horizontal direction. Strata of different permeabilities will give rise to plumes of contaminant with correspondingly different lengths. (The effects of hydrodynamic dispersion have been neglected in preparing the plate representing this process in order to demonstrate this convective effect.)

3

Spreading as a Fluid Phase in Porous Media

3.1 PRELIMINARY EXPERIMENTS

The purpose of this work was to test whether theoretical considerations will permit the prediction of the behavior observed in idealized model systems, i.e., in experiments that simulate nature on a small scale. The physical models used included troughs and columns.

Preliminary experiments with glass columns were carried out first since they promised useful information and required only a small investment in time. All of the columns were composed of Duran borosilicate glass (similar to Pyrex) from Schott and Assoc. (Mainz, West Germany). The dimensions of the columns used were from 20 to 40 cm in diameter and from 100 to 200 cm in height. The porous media used were precision-graded quartz sands manufactured by V. Busch, Schnaittenbach/Opf. These sands are very uniform (Figure 1). This characteristic is required for a large hydraulic conductivity. It also allows a rather homogeneous packing of a column to be achieved, provided the necessary care is given. The light colors of the sands contrasted strongly with the CHC since the latter was marked with red dye at low concentrations (1–2 g/L). In addition, the boundary between the sand that was unsaturated with water and that which was saturated was easily distinguished. These visual contrasts were important since the progression of the CHC flow process during the spills was to be followed photographically. Figure Va illustrates three 1-m-high, 20-cm-diameter

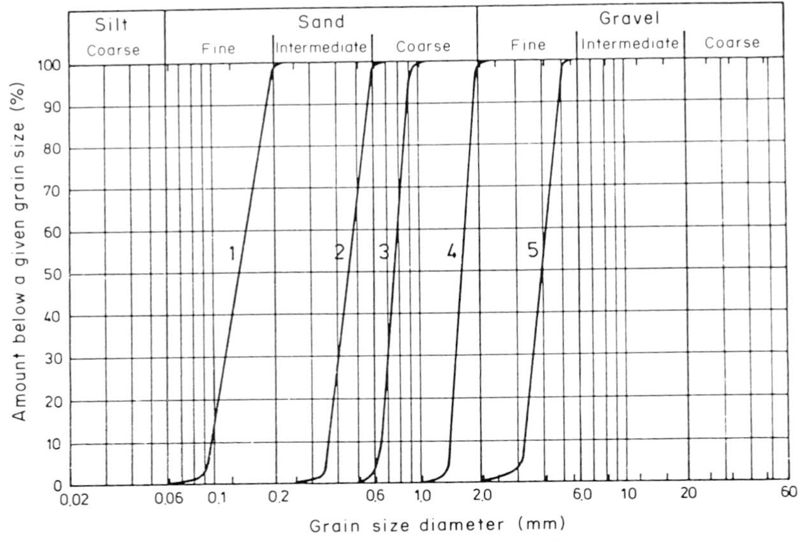

Figure 1. Size distribution curves for model sands.

columns in which it was possible to determine: 1) the infiltration velocity in porous media of different permeabilities; and 2) the retention capacity observed after a spill. The columns in Figures Vb, VIa, and VIb have diameters of 40 cm.

The first experiments demonstrated that CHCs will infiltrate into dry porous media noticeably faster than will water. This was fully expected in view of the lower kinematic viscosities of the CHCs. As a CHC penetrates deeper, the successive leading fronts will not be flat: protuberances will form relatively quickly on those fronts. The protuberances will form more readily when the column diameter is large. When the capillary fringe is reached, it will obstruct the entry of a CHC into the saturated zone. However, whenever there is sufficient CHC supply and pressure, the capillary fringe will not be able to prevent that entry. When the permeability of the saturated zone is high, the sinking of the CHC through that zone will usually proceed remarkably quickly.

In media of low permeabilities (i.e., somewhat lower than

10^{-4} m/sec) the saturated zone will already offer substantial resistance. If the supply of CHC is low, this can bring the sinking of the CHC to a standstill. The pore diameters in this case are so narrow that a large pressure is required to: 1) overcome the interfacial tension between the CHC and water; and thereby 2) break the hypothetical film around the CHC drops as the CHC is forced into the water-filled pores.

In order to demonstrate the effect of the permeability of the porous medium on the sinking of trichloroethylene (TCE) into a groundwater aquifer, three columns were filled with sands of differing hydraulic conductivity (Figure 2). The groundwater table was set at only 10 cm beneath the upper filter plates. A 2.0-L volume of TCE was then added over a time period of a few minutes to the upper filter plate of each of these columns. In column III, the TCE broke

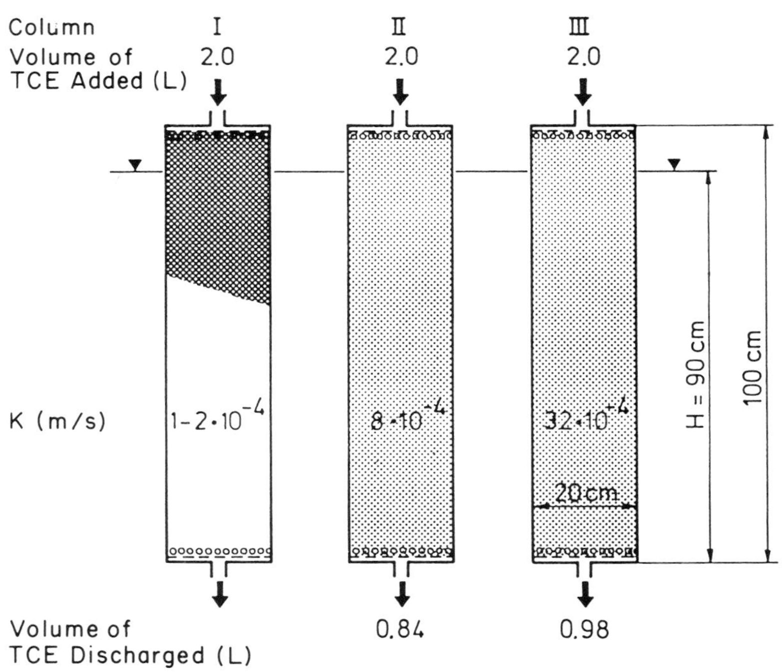

Figure 2. Sinking of TCE in sands of different permeabilities.

through the lower filter plate after approximately 20 min. In column II, the TCE broke through after 60 min. After about one day, the amounts of TCE discharged out of columns III and II were 0.98 and 0.84 L, respectively. In contrast, after one day, the TCE in column I had only penetrated to 30 cm below the water table.

3.2 TROUGH EXPERIMENTS

After orienting ourselves with the above column experiments, we took the risk of carrying out scaled-up experiments that examined the spreading of CHC in our largest available model trough. After tedious preparations, this trough (560 × 160 × 28 cm) (originally designed for groundwater experiments) was successfully outfitted for safe experiments with solvents (Figure 3). The model groundwater aquifer was constructed with two strata. The ratio between the K values of the two strata was 1:4. The thickness of the upper layer was 90 cm, and its K was 0.8×10^{-3} m/sec. The thickness of the lower layer was 60 cm, and its K was 3.2×10^{-3} m/sec. The sands were introduced into the trough in 1-cm-thick layers that were strictly parallel to the bottom of the trough. The gradient (i = 0.005) was produced by tipping the trough. In this manner, the water table and the bottom of the

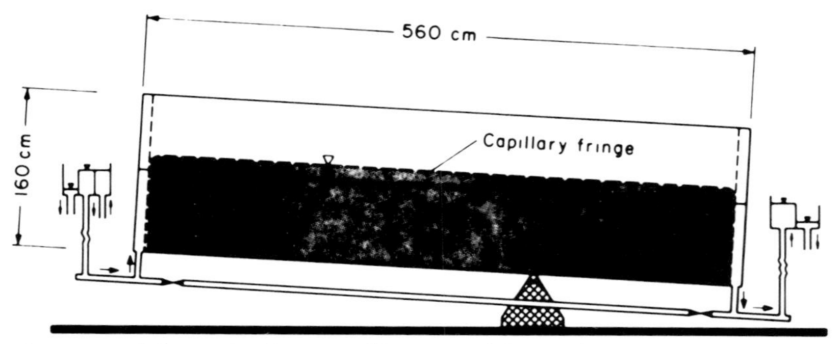

Figure 3. Schematic diagram of large model trough.

trough were parallel. One of the side walls of the trough was constructed of 20-mm-thick safety glass. The course of the spreading in each experiment could then be followed visually and recorded photographically. Perchloroethylene (tetrachloroethylene, or PER) was chosen as the experimental solvent since it possesses the lowest vapor pressure, the slowest vaporization rate, and thus the largest vaporization number of the CHC compound class. The room temperature was 20 to 22°C.

So as to allow a comparison of the results to be obtained with those observed with water, water colored with 1 to 2 g/L of dye was first infiltrated under otherwise equal hydraulic conditions (Figure 4). (The PER infiltrated in the subsequent experiments was also colored with 1–2 g/L of dye.) After 20 min, the solution reached the upper boundary of the capillary fringe. The solution penetrated into the aquifer to a depth that was consistent with the magnitude of the head. However, because of its small density, the dyed solution was forced very quickly in the horizontal direction of flow. This characteristic of the movement of the dyed water was determined by the stratified nature of the aquifer.

After a thorough rinsing of the model trough, 24 L of PER was introduced over 2 hr. The purpose of this experiment was to simulate a leak from an underground pipe (Experiment I). The individual stages of the spreading are presented in Figures VIIa–IXa and 5.

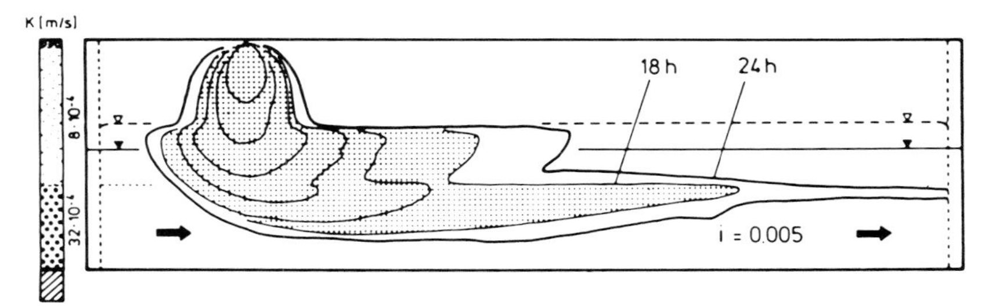

Figure 4. Spreading of the dyed water solution in the two-layered groundwater aquifer in large model trough. Infiltration rate = 15.3 L/hr.

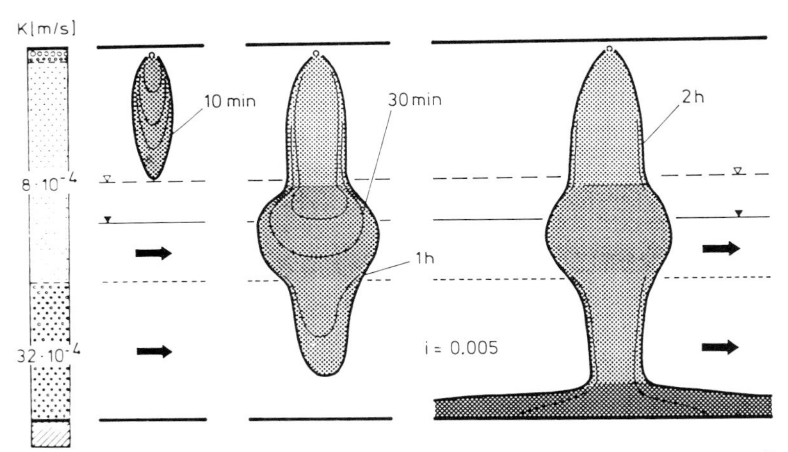

Figure 5. Spreading of PER in large trough (Experiment I, intermediate stages). Infiltration rate = 12 L/hr.

In comparison to the dye plume obtained with water, the body of the PER plume in the unsaturated zone was narrow, almost cigar-shaped. The PER reached the upper boundary of the capillary fringe in only 10 min. The capillary fringe retarded the infiltration of the PER somewhat. It also caused a moderate lateral spreading. Nevertheless, the PER then penetrated into the aquifer rather quickly. As the PER moved into the more permeable layer, a narrowing of the spill body occurred. In just under 1 hr, the bottom of the trough was reached where the PER then formed a flat, watch glass--shaped mound (maximum height about 12 cm) that extended in the direction of the flow (Figure IXa). (In general, though, a CHC will follow the slope of the impermeable layer even when it is in a direction different from that of the gradient of the hydraulic head.) It remained in close contact with the bottom of the trough, demonstrating that the relief on the bottom of an aquifer will alone determine the character of this portion of the spreading (Figure 6).

In just under 2 hr, PER began to flow out from the porous medium on the left side of the trough and into the *upgradient* reservoir chamber (i.e., into the square cross-section cham-

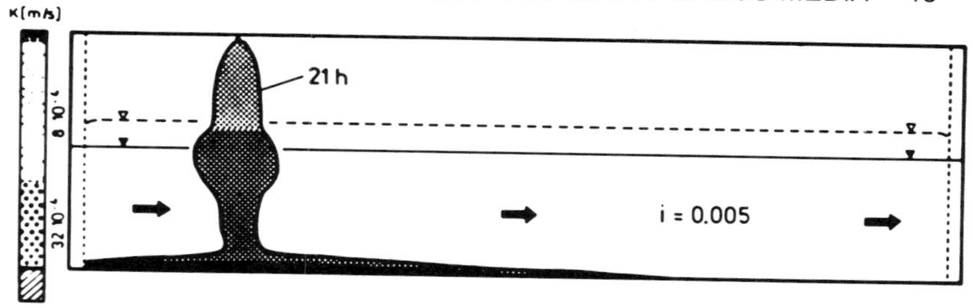

Figure 6. Spreading of PER in large trough (Experiment I, final stage).

ber between the leftmost wall of the trough and the porous medium). The thickness of the zone of PER on the plane of the chamber/porous medium interface was only about 3 cm. The passage of the PER from the porous medium into the chamber occurred as a series of drops that appeared like a string of pearls. The drops jostled against one another but did not flow into one another. They dropped into the pipe supplying water to the chamber. Seven hours after the beginning of the experiment, a total of 8 L was recovered from the pipe, thereby accounting for about a third of the PER originally added. After a total of 21 hr, the spreading of the PER ceased.

For Experiment II, the water table was lowered to be within the lower, more permeable stratum. The same hydraulic gradient was maintained (Figure 7). The purpose of this experiment was to simulate a sheetlike spill such as might occur with a tank car accident. Two watering cans were used to apply 36.3 L of PER to a 100-cm-long section of

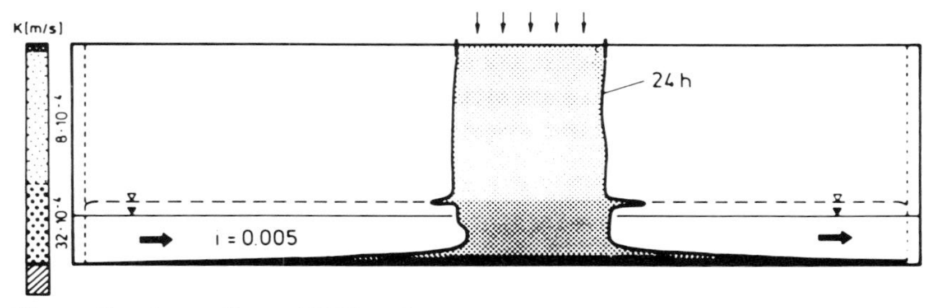

Figure 7. Spreading of PER in large trough (Experiment II, final stage).

the sand at the top of the trough. The application was carried out over 4 min. This corresponded to an infiltration rate of 2 $m^3/hr\text{-}m^2$. This led to a nonuniform but not unrealistic type of application. Individual strands of PER penetrated downwards very quickly, and within 5 min reached the upper boundary of the capillary fringe (Figures Xa and Xb). After approximately 15 min, the infiltration front was fully closed, and the strands appeared like tassels on the fringe of a rug. The width of the infiltrating mass was equal to the width of the spill at the surface. The infiltrating mass exhibited an almost constant cross section throughout the depth of the spill. A flat mound of PER started to accumulate on the bottom of the trough only 10 min after the beginning of the experiment (Figure Xb). After about 1 hr, PER began to appear as a steady series of drops entering the downgradient reservoir chamber (i.e., the square cross-sectional chamber between the rightmost wall of the trough and the porous medium). After 4.5 hr, small amounts began flowing into the upgradient reservoir chamber. After 5 hr, the amount of PER that had been recovered was 9 L. The larger application rate and area for this experiment allowed the infiltration to occur substantially more quickly than in Experiment I.

An important result of these experiments is the conclusion that permeable aquifers ($K >> 10^{-4}$ m/sec) will apparently be penetrated quickly whenever the CHC infiltration rate is adequate. Under such conditions, the bottom of the aquifer will be the controlling spreading surface. The groundwater flow (with $i = 0.005$ as in the experiment) will not cause a deformation or deflection of the CHC mass even after a prolonged period of time. The figures showing the final distribution of the PER are worth noting: the amount of PER in the unsaturated zone is substantially less than in the saturated zone. The figures show "bled-out" CHC masses in which the only CHC that remains is in the state of residual saturation. The only place that high degrees of saturation exist is at the bottom of the aquifer. The CHC located there will still be mobile, however, and could be partially recovered with a well installed in a sump placed in the confining layer (Figure 8).

The infiltration behaviors of two different CHCs (dichloromethane [DCM] and PER) were compared in a glass-walled trough that was shorter and higher than the large trough but could not be placed on an incline (Figures XIa–XIIb and 9). The hydraulic conductivity of the sand used was approximately 1 to 2 × 10^{-4} m/sec. The porosity was 50%. The progress of the infiltration could be followed by both movie and still cameras on both sides of the trough. Initially, the sand was completely saturated. The water was lowered to the selected level 12 days before the start of the experiment; the unsaturated zone was thus moist for the experiments.

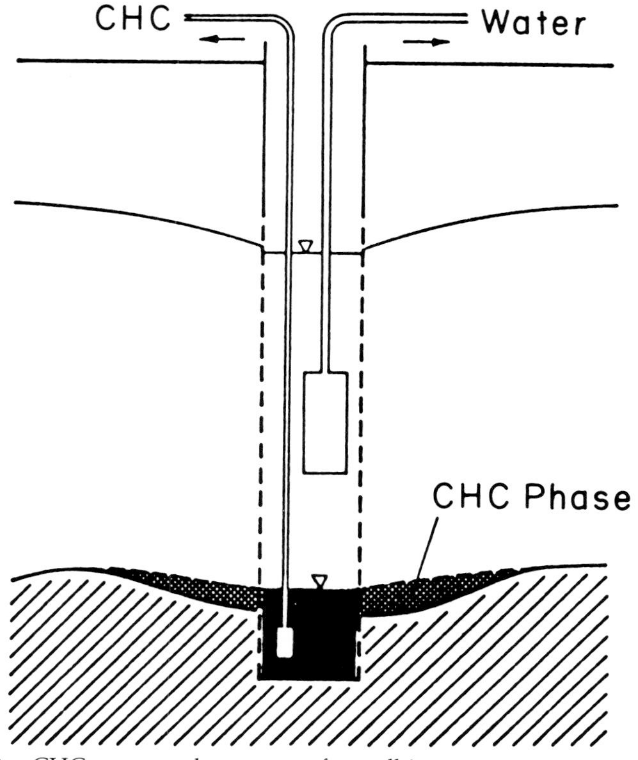

Figure 8. CHC recovery by means of a well in a two-pump system.

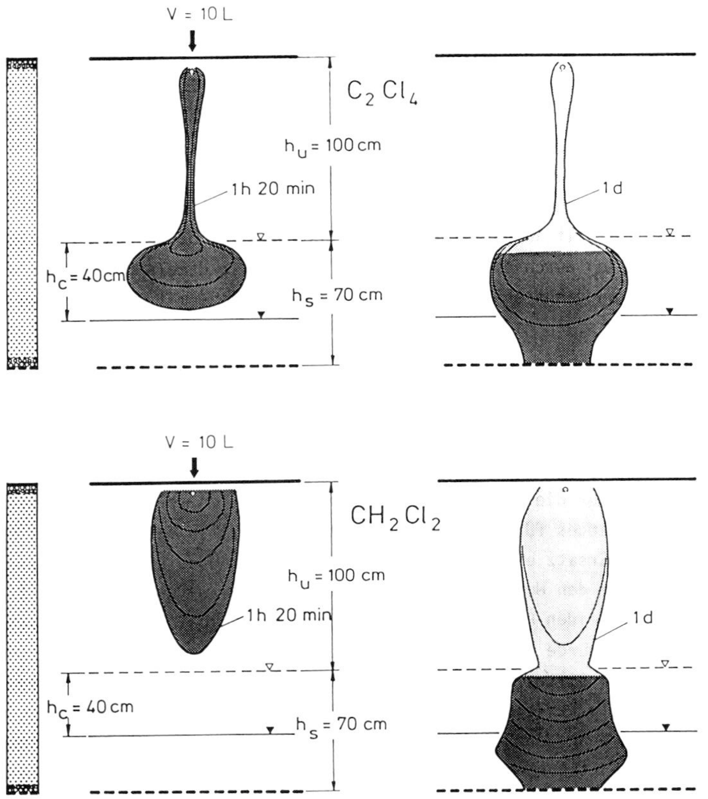

Figure 9. Glass trough; groundwater with zero gradient. K = 1–2 x 10^{-4} m/sec. Upper figures: PER. Lower figures: DCM. Right figures: final stage of infiltration.

Under otherwise identical conditions, the infiltration of DCM occurred both more broadly and more slowly than did PER. The breakthrough of PER to the bottom plate occurred after 4 hr. For DCM it occurred only after 10 hr. On the basis of the lower kinematic viscosity of DCM (ν = 0.32 mm^2/sec) one would have expected a quicker and more narrow penetration for DCM than for PER (ν = 0.54 mm^2/sec). The higher vapor pressure and vaporization rate of DCM are probably responsible for this result. [Translator's Note: These and other physical data may be found in Appendix I.] After reaching the saturated region (capillary fringe), their effects cease.

Figures VIIa-Xb were obtained during the experiment that we documented by movie camera. This is pointed out for those readers who have seen that film. Conversations with those who have seen the film have revealed that the film gives the impression that CHCs are fundamentally capable of: 1) penetrating quickly into both the unsaturated and saturated zones; and 2) penetrating all the way to the bottom of the aquifer when a significant supply of CHC is available from above. These generalized perceptions require a qualification: in the saturated zone they apply only to the more permeable types of soils. At the time that film was presented, we were unable to make any comments for soils that are only slightly permeable. These experiments have, in the meantime, been performed.

3.3 COLUMN EXPERIMENTS WITH LOW-PERMEABILITY SATURATED MEDIA

In order that the results of these experiments could be compared most easily with those obtained with the higher permeability media, we would have liked to use a similar model trough. Because of the large amount of work involved in the excavation and refilling of the large trough, however, we chose to use a 100-cm-high, 40-cm-diameter glass column (Figure Vb). For the test sand, we chose Model Sand 1 ($K = 1 - 2 \times 10^{-4}$ m/sec). The two filter regions (i.e., sand packs) were constructed as mirror images of one another (Figure 10). The test sand was painstakingly added in horizontal layers in order to prevent the infiltrating fluid from moving down any inclined strata or other preferential pathways formed during the packing. The water table was raised above the upper boundary of the sand filter layer so that the injection needles dipped into the water. This allowed the fluid to be added to the column slowly and in the most

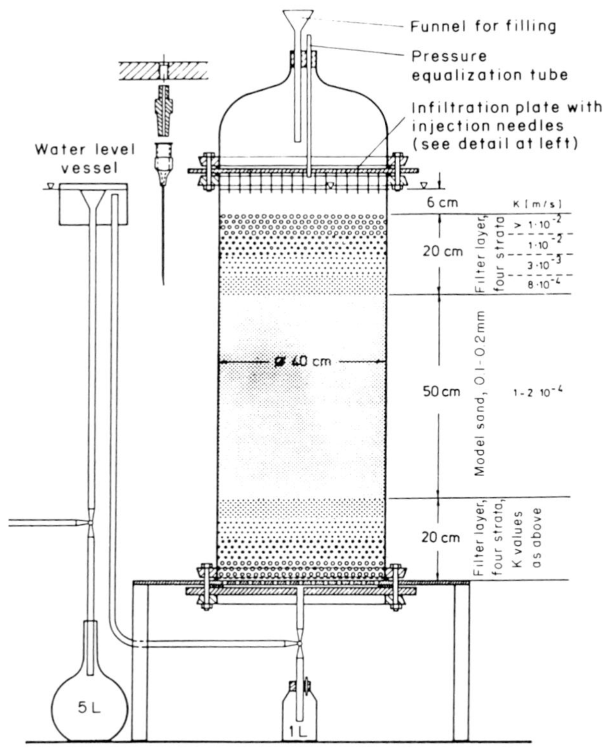

Figure 10. Column experiment; saturated porous medium with low permeability.

uniform manner possible. This experimental setup was used to a certain extent as a prototype for a whole series of experiments for testing a wide range of sands.

The first 5 L of PER was added to the column in just under 0.5 hr. The PER penetrated to the boundary between filter layers 1 and 2. The PER then spread laterally. Breakthrough to the boundary between layers 2 and 3 followed, and this was in turn followed by a breakthrough to the boundary between layers 3 and 4. The upper boundary of the test sand was then reached. Only slowly did the individual filter layers become saturated with PER. To our surprise, the PER did not penetrate into the test sand. After 3 days, 2 L of PER were added so that the total amount of PER added became 7 L. Even after this addition, the character of the PER front did not change in the least. After 14 days, we started to become impatient. Since it appeared that no change could be expected unless something substantial was done to the system, we lowered the water level vessel by 20 cm. (Up until that time, a total of only 3.25 L of water had been drawn off the system.) The PER then penetrated into the test sand. The front was not uniform, nor even bulb-shaped; rather, the PER penetrated as tendrils or fingers in the interior of the sand body. After about 2 hr, approximately 2 L of PER had passed out of the column. The experiment was ended at this point.

A new experiment was defined and planned in which the water table would be lowered slowly in small increments. In other words, the head difference between the column and the regulating water vessel would be increased in small increments. A 20-cm-thick layer of the test sand (Sand 1 again) was placed in a column 50 cm high. This time, the test sand was underlain as well as overlain by only two filter layers. The upper filter layers were the same as the two uppermost layers in Figure 10. The lower filter layers were the same as the two lowermost layers in Figure 10. The initial water level was set at 5 cm above the sand body. The addition of PER proceeded as in the previous experiment. After the addition of PER was completed, however,the bell jar with the infiltration plate was removed, and a water level point gauge was mounted on the column. (A point gauge is a movable steel needle capable of making an electrical circuit [and lighting a bulb] when it touches the water table.) The gauge mounted on the column is visible above the number *4* in Figure VIa. A point gauge was also mounted on the water level vessel. In this manner, the difference between the two water tables could be measured exactly.

The addition of the PER (5 L) was carried out by dripping it into the water overlying the sand in the column over a period of 20 min. The penetration of the PER into the upper filter layers was again quite slow, and the PER again came to a standstill when it reached the upper surface of the test sand. At that point, the two upper filter layers were about 60% saturated with PER, and the remaining PER formed a 2-cm-thick layer on top of the upper filter layer. It should be noted that 17 hr after the beginning of the experiment, only 2.7 L of water had been forced out of the system. Since the amount of PER added was 5 L, the water level in the column was just under 2 cm higher than in the water level vessel ($3.14 \times [20\ \text{cm}]^2 \times 1.83\ \text{cm} = 2300\ \text{mL} = 5000\ \text{mL} - 2700\ \text{mL}$). In spite of the overpressure in the column, the PER did not penetrate into the test sand. This gave the impression that the PER was resting on an impermeable membrane.

After 7 days, we began to lower the water level vessel. It is important to note that this point in time was taken as time zero for the rest of the experiment. The measured values of the experimental parameters referenced to that time zero are given in Figure 11. The water table in the column reflected the sinking only slightly. Although the thickness of the PER layer above the upper filter layer decreased prior to day 11, no PER entered the test sand. The additional PER that moved into the filter layers only increased the degree of PER saturation in those layers. On day 11, the PER began to enter the test sand. After the water table difference reached 13 cm on day 25, PER began to flow from the bottom of the column for the first time: the PER had begun to break through from

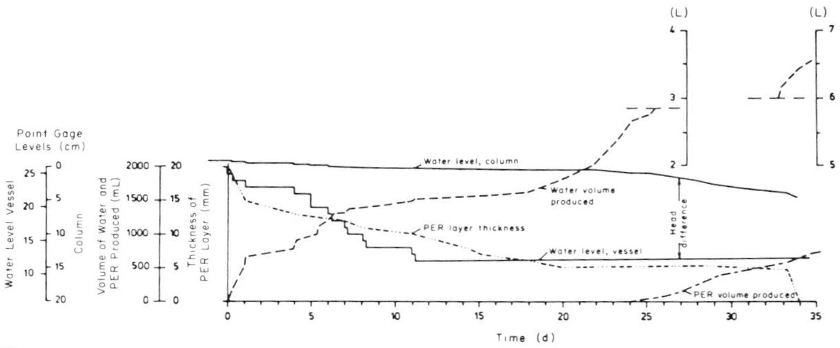

Figure 11. Results obtained in Figure VIa experiment.

the column. Prior to this, red spots and streaks had begun to show on the glass walls of the column in the test sand zone. Thus, the breakthrough occurred *in the interior* of the column, and wall effects definitely played no role; PER that reached the walls did so from the interior of the column. On day 34, the PER that remained on the top of the upper filter layer disappeared together with the overlying water. Figure VIa, which is a photo taken near the end of the experiment, shows that a uniform distribution of PER still did not exist in the test sand at that time.

Although we cannot give final overall interpretations for these two experiments, they have produced an important preliminary result. Even when we take into consideration that results obtained with a column (which has lateral boundaries) and those obtained with a trough (which has no effective lateral boundaries) are not directly comparable, there is nevertheless no doubt that a substantial fluid pressure is required to drive an immiscible fluid into and through a water-saturated, fine-grained layer. To be sure, the two experiments described in Figures 10 and 11 do not allow a specific determination of the magnitude of the pressure that is required. They nevertheless substantiate the earlier observations that for strata with hydraulic conductivities less than about 10^{-4} m/sec, CHCs will often not be able to penetrate (i.e., enter) at all, or will only penetrate very slowly. The only condition under which penetration will occur into a medium of this type of permeability is when the CHC exerts a relatively high pressure; this can occur when high application rates are concentrated over a limited area as in the experiment described in Figure 9.

We have thus refuted the occasionally expressed, generalized perception that the dense nature of the CHCs (relative to water) will always allow them to sink into almost any type of system provided that the supply of CHC is continuous. This can be the case in the unsaturated zone where the high density of a CHC is fully effective. In the saturated zone, however, the effective density is about 1 g/mL less due to buoyancy effects, and the driving force is decreased accordingly.

3.4 GLASS FRIT EXPERIMENTS

We have sought to develop a simple and easily carried out procedure for determining the relationship between the size of the pores of a fine-grained medium and the magnitude of the pressure that is required for the penetration of CHCs. This development has led to the use of glass frits (glass filters) with exactly determined pore sizes in place of different types of soils. Glass frits with a diameter of 9 cm and thickness of 0.9 cm were fused into 26-cm-long glass columns (Figure 12). Such columns can be used with either model sands or glass beads, or without them, as desired. The adjustment of the water level vessel allows the creation of any desired pressure difference.

The filters used, listed in Table 1, were made from Duran borosilicate glass obtained from Schott and Assoc. (Mainz, West Germany). The pore sizes specified refer to the largest pores of the disks. Each specification thereby defines the minimum diameter of the particles (in this case, droplets) that will not pass through that frit size without first being decreased by deformation.

In one of the first series of experiments, the columns were filled with water to a level of 10 cm above the frits. (No packing was used in the columns.) The air bubbles in the frits were removed by suction with a tube. Then, 65 mL of dyed (red) PER was added directly over the frit with a perforated tube. The drops gathered on top of the frit, and in a manner similar to the behavior of drops of mercury, did not wet the frit. The drops ultimately flowed into one another and formed a 1-cm-thick layer of PER. The water column was thereby raised by 1 cm. In no case did the PER immediately penetrate into the frit. The water level adjustment vessel was then lowered in a stepwise fashion, and the head difference

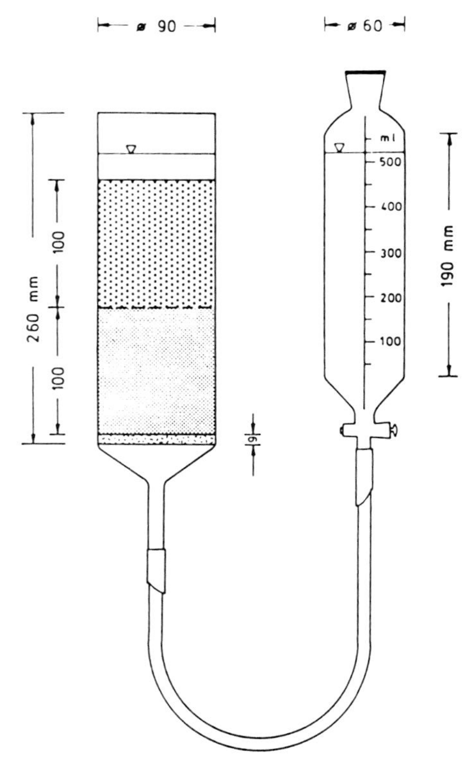

Figure 12. Diagrammatic sketch of a glass frit column with water level adjustment vessel. The column can be used without any layers of sand or glass bead packing (shown as stippled regions) or with such layers.

Table 1. Pore Sizes of Glass Filters Used in Glass Frit Experiments

Porosity	Symbol	Maximum Pore Size (μm)
00	–	250 to 500
0	P 250	160 to 250
1	P 160	100 to 160
2	P 100	40 to 100
3	P 40	16 to 40

between the vessel and the column (Δh) measured. The point at which breakthrough occurred on the bottom of the frit was determined.

With the 00 frit, the PER began to break through 1 hr after it was added. The breakthrough occurred even before the water level was lowered and was limited to a single point on the frit. A PER drop fell every 5 sec. It is possible that this pointwise breakthrough was due to a pore that was larger than the others in the frit. The lowering of the water level by 1 cm accelerated the rate of drop formation at that point but did not initiate drop formation elsewhere. Only after the water level was lowered by a total of 3 cm did breakthrough begin simultaneously at six new places on the frit. The water level had to be lowered by a total of 12 cm before breakthrough took place over the entire surface of the frit. After a majority of the PER had flowed through the frit, a small portion of the frit was freed of PER. Prior to this point in time, the amount of water in the cylinder had been constant. Once the hole appeared in the PER layer, however, the water flowed through quickly. After a short period of time, all but a small amount of the PER disappeared through the frit.

With the frit of pore size 0, the PER began to break through at numerous places when $\Delta h = 7$ cm. One could observe a virtual rain of PER drops from the underside of the frit. With the frit of pore size 1, the breakthrough also occurred over the entire surface of the frit, but in this case the Δh required was 10 cm. The small droplets of PER were again rainlike in appearance. With the frit of pore size 2, the Δh required was 20 cm, but the PER passed through only in the middle of the frit and only at a few places. With pore size 3, $\Delta h = 45$ cm failed to initiate breakthrough. When the water level adjustment vessel was raised and then relowered, an air bubble appeared under the frit. At this point, a strong raining of PER droplets began. (It is possible that the air bubble facilitated the initial breakthrough of PER.) When the water level vessel was again raised so that $\Delta h = 26$ cm, the rain of PER droplets decreased in intensity but, nevertheless, continued to flow.

Although these few experiments were not comprehensive, their results reveal a very plausible relationship between Δh for the beginning of breakthrough and the pore size of the frit (Figure 13). This relationship shows that the pressure required to penetrate water-saturated pores increases very rapidly as the pore size decreases below 0.5 mm (500 μm).

Subsequent to this work, experiments in which the columns were filled with sand were also carried out. While it was not possible to conduct this work in a completely systematic manner, the results obtained substantiated the previous observations. In particular, they demonstrated that even in a porous medium in which a CHC from a simulated leak has quickly penetrated through an aquifer (e.g., the experiments in Figures VIIa-VIIIb), no penetration of a CHC through a frit will occur when the layer of CHC accumulated on that frit is thin; a certain CHC pressure must be developed before the CHC can move from a stratum with large pores and penetrate into a next deeper stratum with smaller pores.

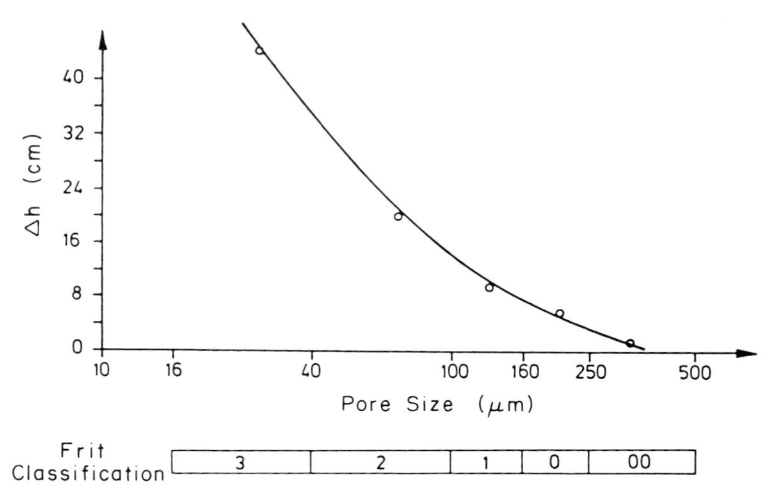

Figure 13. Relationship between the pressure required for the beginning of breakthrough of PER through water-saturated glass frits and the frit pore size.

The conclusions regarding the overall behavior of CHCs in the saturated zone will be discussed later. Nevertheless, it should be pointed out now that CHCs which have penetrated into a groundwater system will not be able to penetrate further by virtue of a vertical gradient in the *water* head. Rather, a CHC will only be able to penetrate because its density is greater than the 1.0 g/mL of water. It is also important to note here that the porous media of frits is characterized by a pore size and not by a particle size. To the best of our knowledge, a simple relationship between the two parameters does not exist.

3.5 LYSIMETER EXPERIMENT

The process of the infiltration of CHCs into the unsaturated zone must be distinguished sharply from the sinking of CHCs into the saturated zone. In order to complement the experiments carried out with the model sands, we have carried out a lysimeter experiment with a natural porous medium. Our main goal was to learn about the effects of heterogeneity. A 200-cm-thick profile of alluvial sediments was obtained from a gravel pit located in the slope of the lower terrace of the upper (northern) Rhine rift valley. The profile was removed layer by layer from the region that lay beneath the soil zone and above the upper boundary of the capillary fringe. The profile was reconstructed layer by layer in a 200-cm-high, 40-cm-diameter glass column.

The profile comprised primarily intermediate to coarse sands, as well as fine gravels. As a group, such materials are considered rather permeable. The particle size distribution data are presented in Figure 14. The 95 to 100-cm interval contained a large portion of fine sand. This layer was cemented by ferriferous carbonates. After the column was filled, it was flooded from the bottom to the top. Thereafter, the column was allowed to dewater for 6 days. The resulting water content represented approximately 20% of the pore

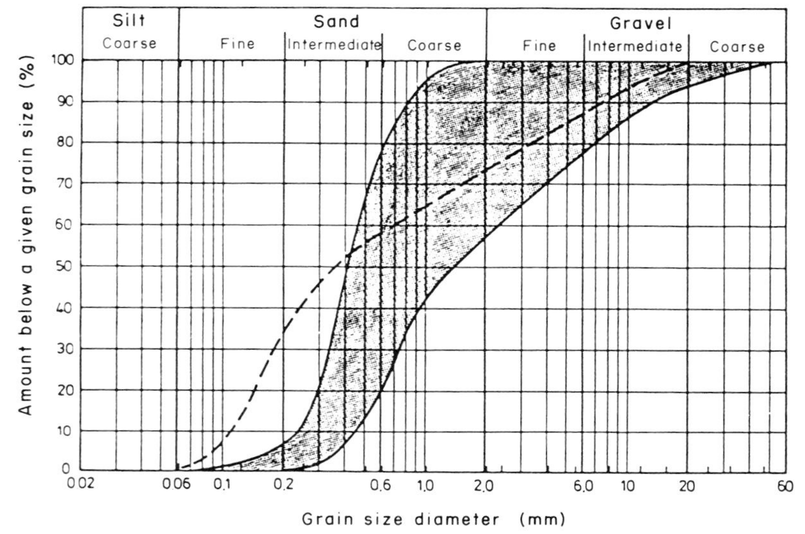

Figure 14. Lysimeter experiment; size distribution data. The dashed curve is the size distribution for the material in the thin layer between 95 and 100 cm. The size distributions for all of the other layers lie between the two solid curves.

volume. PER was selected as the representative CHC for this column for the previously mentioned reasons.

Before adding PER to the column, however, we first desired to investigate the behavior of the column with respect to water infiltration. To this end, for a period of 21 days, two applications of 10 mm (1.4 L) of fluorescein-dyed water were made each day to the top of the column. The progress of the dye front through the column was studied with an ultraviolet lamp. The positions of the dye front at successive times during the course of the experiment are given in Figure 15. As Figure 15 shows, the infiltration proceeded very irregularly. The front formed was rather distended in places. After only 22 hr, a broad tongue had formed. This tongue proceeded through the column much more quickly than did the front in other portions of the

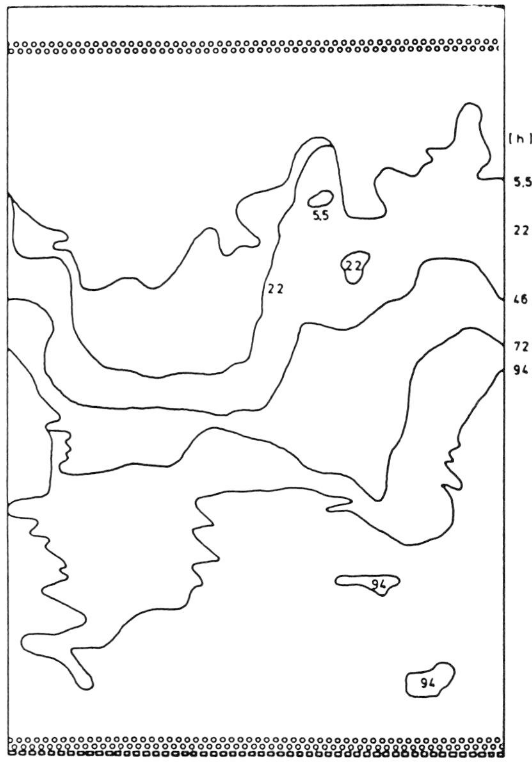

Figure 15. Lysimeter experiment; development of the infiltration front for simulated fluorescein-dyed "precipitation."

column cross section. After 94 hr, the tongue had almost reached the lower filter layer.

After the fluorescein-dyed water had been rinsed from the column, the column was flooded again and drained for 10 days. The addition of the PER then began in aliquots. Within 8 days, a total of 14 L of PER had been added (Figure 16). The downward progress of the PER proceeded in a manner that was even more irregular than that observed with water (Figure 17). The intensive application of PER gradually led to a complete saturation of the upper 50 cm of the column. This upper 50 cm began to slowly "bleed out" PER from within.

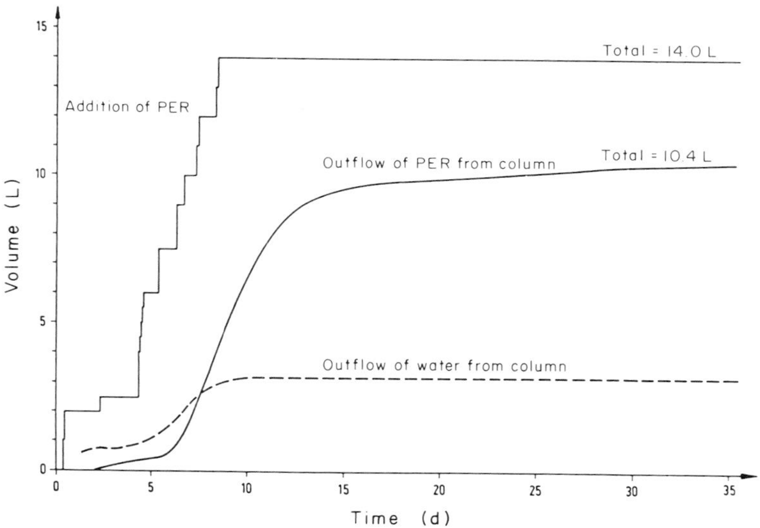

Figure 16. Lysimeter experiment; curve for addition of PER, and curves for the outflow of PER and water.

The further addition of PER led to a flow of PER down through the interior of the lower portion of the column. This probably proceeded through channels which were prewetted with PER. It is also probable that some lateral spreading of PER took place on the fine-grained sand layers. Indeed, evidence for such lateral spreading was found in that PER appeared at a few places on the glass walls.

The greatest accumulation of PER occurred on the top of the fine-grained, carbonate-containing, 95- to 100-cm layer. This temporary accumulation reached a maximum thickness of approximately 10 cm. The PER penetrated this less permeable layer only rather slowly and only through a few "weak points." The further infiltration proceeded by means of single fingers of solvent within the interior body of the sand. When the fingers reached some of the layers, lateral spreading occurred and allowed some PER to reach the glass walls. A complete saturation of the sand body did not occur. Individual fingers reached the bottom of the lysimeter after only 2 days, whereupon PER began to drip out. After 36 days, a

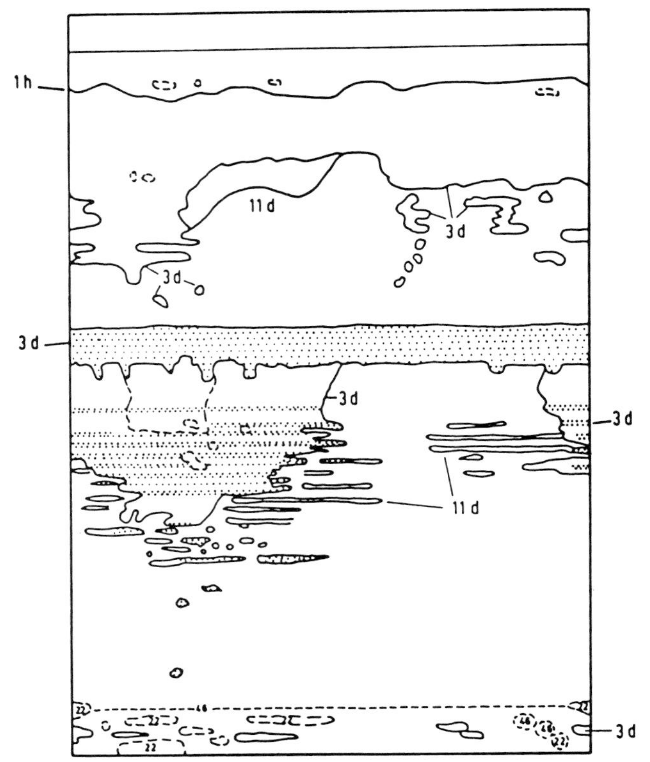

Figure 17. Lysimeter experiment; development of the infiltration front for PER.

total of approximately 3.2 L of water and 10.4 L of PER had flowed out of the column. Therefore, a total of approximately 3.6 L of PER was retained in the column. This indicates that the overall PER retention capacity of the column was 15 L/m^3. [Translator's note: This retention capacity value averages the PER content of those regions containing PER together with those not containing PER.]

In cases involving spills of solvent, there arises the fundamental question whether it will be possible to displace a CHC phase out of the unsaturated zone to the saturated zone, from where it could be recovered by a well. Such infiltration could occur either by infiltration of water with a sprinkler system, or with drain piping located above or below ground. However, theoretical considerations indicate that only minimal success can be expected in such attempts. In fact, they will tend to extend and deepen the zone in which the CHC is held. Thus, while some small scale mobility may be imparted to the CHC, it will not flow in any substantial manner to a well as a pure phase. Rather, the major movement will be in the dissolved form. Indeed, treatment with water can be effective in certain circumstances in washing out the CHC by dissolution. In this dissolved form, one can recover the CHC in a known manner.

The theoretical considerations discussed above required verification. For this purpose, the PER-containing column described above was then infiltrated with water. For the first two days, the equivalent of 60 mm of rain (7.5 L) was applied each day (three applications of 20 mm [2.5 L] each) to the upper surface of the sand. On the third day, 120 mm (15.1 L) was applied, of which 100 mm (12.6 L) was added in one increment (Figure 18). During the 100-mm (12.6-L) application, the PER was mobilized in certain regions of the column. This was recognized by an increase of about 2 cm in the amount of PER over the 95 to 100-cm layer. On the fourth day of the experiment, a massive, prolonged application of 200 mm (25.1 L) of water was made. This led to a layered increase of PER in the region of the capillary fringe (located near the top of the bottom plate of the column). A total of only 160 mL of PER was recovered as outflow from the column after 45 days. Therefore, of the original 3.6 L of PER present in the sand at the beginning of the experiment, nearly 3.44 L remained, corresponding to a retention capacity of 14.3 L/m^3.

The manner in which the PER was distributed in the column was determined by an excavation of the contents of the column (Figure 19). Only in the upper 80 cm of the column was the PER found to be more or less evenly distributed over the cross section of the column. The breakthrough of PER across the fine-grained carbonate-containing layer

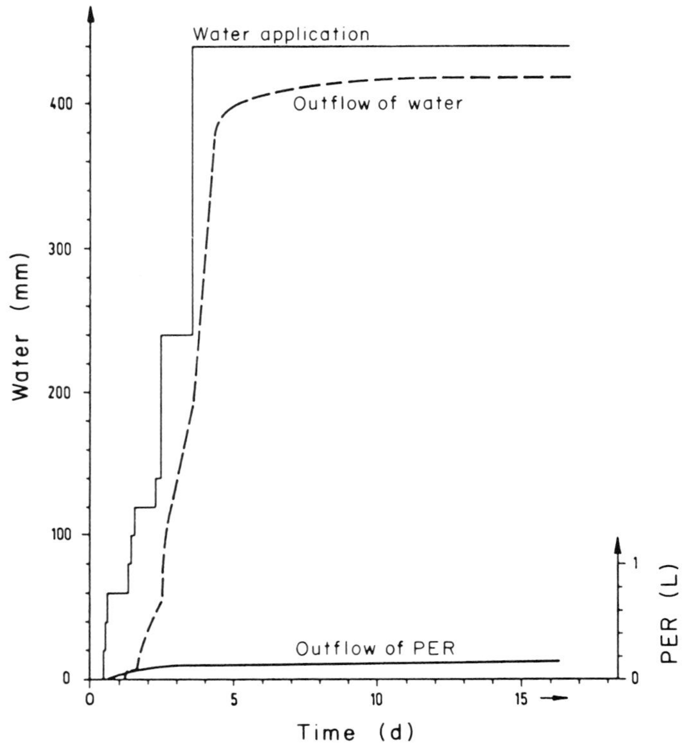

Figure 18. Lysimeter experiment; water ("precipitation") application and PER outflow.

occurred over a limited cross section, as was indicated by the presence there of only a few large patches of PER (see cross section at 0.95 m depth). These patches revealed the fingers by means of which the PER penetrated toward the bottom of the column and from which it was subsequently able to rebroaden its distribution. The greatest spreading occurred on the top of the capillary fringe (cross section at 1.86 m depth). In contrast, only about half of the cross section was saturated with PER above the capillary fringe, and it is in that context that the aforementioned, low retention capacities of 15.0 and 14.3 L/m^3 can be understood.

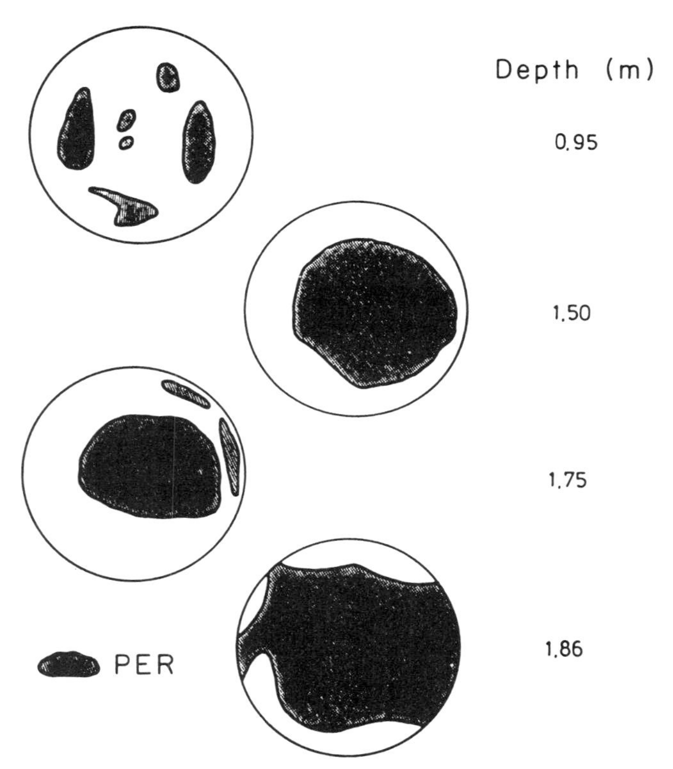

Figure 19. Lysimeter experiment; PER distribution in selected cross sections.

The question is repeatedly asked whether infiltration experiments in glass columns are affected by a preferential downward penetration of fluids along the glass walls. If present, this so-called wall effect has the potential to cause a misinterpretation of the infiltration behavior of a CHC. However, as we have found through the excavation of numerous columns into which CHC infiltration has been carried out, this has either not been a problem at all, or at most, a very minimal problem with the sands used in our work. Figure 19 serves as an example in this regard.

The retention capacity of the filled column is given in Figure 20 as a function of time. The excess PER passed en masse through the column very quickly. After approximately 6 days, the experiment entered into a phase wherein the PER was dripping only slowly from the column. During this period, the PER content of the lysimeter decreased only slowly. On this basis, it is apparent that in the case of a large-scale CHC spill on the ground surface, even a partially effective recovery of the CHC will be possible only if the spill area is excavated very quickly. In addition, it may be noted that the infiltration of the water had very little effect on the final draining of the column.

It is of interest to compare the infiltration of the dyed water (Figure 15) with the infiltration of PER (Figure 17). PER passed through the upper 95 cm of the column more quickly than did water. At the 95 to 100-cm layer, however, the PER was slowed substantially and, in fact, pooled up to some extent. In contrast, the water passed through this level with only little difficulty and continued deeper with an uneven penetration front. Once the PER did penetrate this layer, it suffered further slowing when it reached the capillary fringe; the water passed through the capillary fringe much more

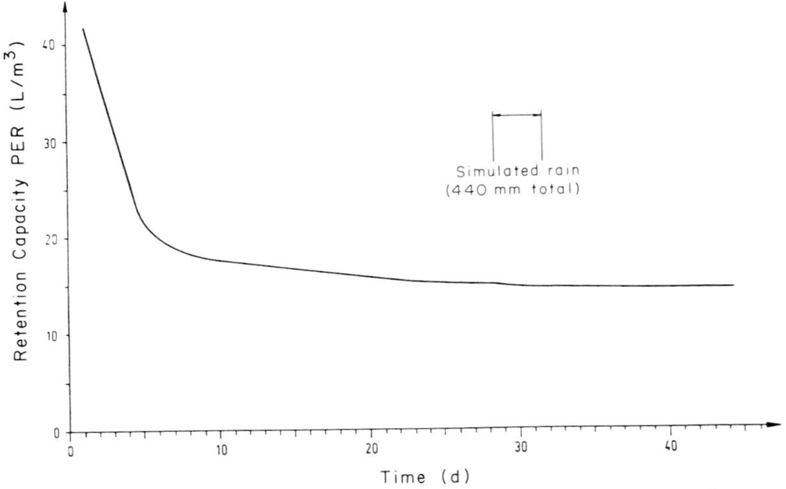

Figure 20. Lysimeter experiment; retention capacity as a function of time.

quickly. In terms of their relative rates of infiltration, PER obviously has the advantage in unsaturated layers due to its high density. Since it is not miscible with water, however, it is slowed in saturated layers and will break through only slowly through the weak points in those layers. Water has a substantial advantage there since it can merely displace downward the water that is already present, thereby effecting a net movement of water through those layers.

It is therefore apparent that the process of CHC infiltration is oversimplified when it is claimed that water-immiscible CHCs will penetrate a formation more quickly than water due to their higher densities and lower viscosities. Where the pore volume is filled with water, a given CHC will only be able to penetrate after it has developed sufficient head of its own to drive out the water. Initially, this will occur at the "weakest" points. With some moist and heterogeneous soils, it is entirely possible that the fluid-mechanical advantages of the CHCs (density and viscosity) will not be effective in causing a rapid penetration of the CHCs.

4

Retention Capacities of Porous Media

As is made apparent by Figures IIIb and IVa, the calculation of the extent of spreading of limited volumes of spilled immiscible fluids in homogeneous media will, in general, depend upon a knowledge of the residual saturation. In heterogeneous systems, there are additional parameters that affect the retention of such fluids. For example, the spilled phase can be concentrated on the surfaces of low-permeability layers, in the capillary fringe, and in depressions on the bottom of the confining layer of the aquifer. The retention capacity of a natural medium is therefore generally greater than a simple residual saturation. Nevertheless, the residual saturations of the layers of a heterogeneous system are the only capacity quantities of that system that can, to some degree, be measured. Therefore, despite the heterogeneous nature of an overall large-scale system (e.g.,layers and laminates), defined residual saturations must be measured in order to have useful numbers to apply to spill problems. One must therefore differentiate between the residual saturation of (1) the unsaturated zone where a chlorinated hydrocarbon (CHC) will find itself wetting the grains (Figure XIVa, CHC in a pendular state); and (2) the saturated zone where a CHC will be present as isolated drops in the open portion of the pore volume (Figure XVa, CHC in an insular state).

It was our goal at this point to obtain an understanding of the order of magnitude of the residual saturation as it depends on the nature of the porous medium. (For expediency, the hydraulic conductivity of the medium was selected as the parameter against which the residual saturation would be measured.) To this end, a standard method for the determination of the residual saturation of the unsaturated zone was developed in stages (Figure 21). Columns with both dry and moist packings were prepared. Excess CHC was then introduced from the top until the porous medium was fully saturated with CHC. CHC was allowed to flow out of the column during this process. The volume of the outflow was measured. For each experiment, the total retention capacity (L/m^3) was obtained (Figure 21a) as the quantity (total inflow – total outflow)/(column volume). It should be pointed out that a "hanging" capillary fringe is always formed above the filter layer/bottom plate of a column. In the case of the experiments to determine the residual saturation of dry packing, the resultant capillary fringe was composed solely of CHC. For the experiments with moist packing, the

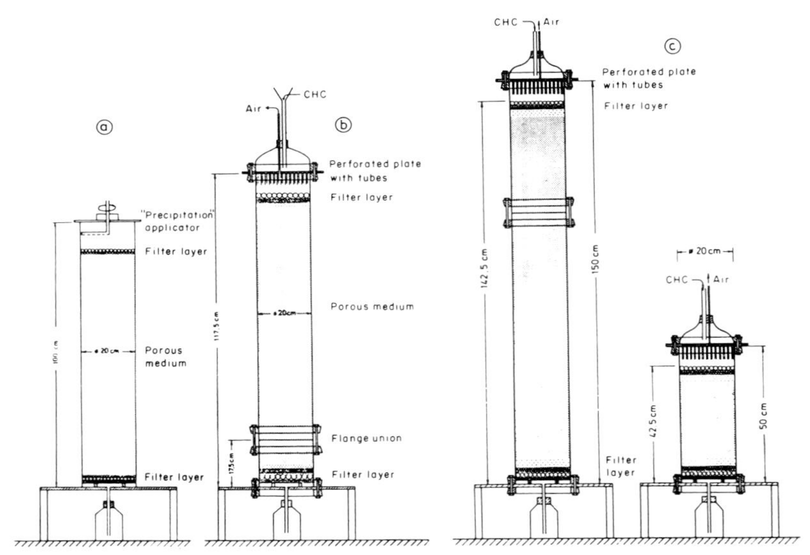

Figure 21. Glass columns for residual saturation determinations. a. One piece column with height of 100 cm. b. Stacked column. c. Column pair.

capillary fringe was initially composed of water; after the addition and draining of the CHC, the capillary fringe became enriched in CHC. Therefore, for both types of columns, the whole-column retention capacity values were always larger than the residual saturations. The error increases as the medium becomes more fine-grained. The percent error can be decreased by using taller columns, but then the expense of the experiments increases.

In order to quantify the magnitude of this effect, we extended the length of the 1.0-m column with a section of column. The latter was always larger than the maximum height of the capillary fringe (Figure 21b). At the conclusion of the draining phase of an experiment, the two sections were separated with a piece of tin sheet metal. The amount of CHC in the smaller section was determined chemically; the result was subtracted from the total amount of CHC retained in the two-section system. Dividing that corrected quantity by the volume of the upper section gave the true retention capacity. However, the chemical determinations were time consuming and often suffered from artifacts due to vaporization losses. We therefore abandoned the use of this type of experiment and switched to using pairs of columns as described in Figure 21c. The smaller column was set up and used separately. The larger column contained a section that was equal in length to the smaller column. Both columns were handled identically. When the height difference between the two columns was taken into consideration, the retention capacities of the two columns led directly to the actual residual saturation. The height difference between the columns was 100 cm. With this method, it was not necessary to separate two columns or to carry out any chemical determinations.

Residual saturation numbers are needed for dry media (as under industrial facilities or warehouses) and for moist media. Therefore, we undertook to determine residual saturation values for both conditions. Moist media were prepared by completely saturating packed columns from below with water. The water table was then lowered to below the level of the bottom plate. The field capacity was reached within a few days. Columns in this condition served as the reference or standard for moist media. Figure 22 shows how the dewatering phase progressed for the sands used here. In order to obtain a uniform distribution of CHC in a column, the CHC was introduced at the top with small injection needles (held in place in a perforated plate) until the column was uniformly and completely impregnated. In order to avoid vaporization losses during the drainage phase, the solvent leaving the column was collected under a layer of water in a vessel.

The downward progress of the CHC phase front as visualized at the glass wall was recorded on the wall with a felt pen. These plots were then transferred by tracing them onto transparent paper (Figure 23). In this manner, the formation of fingers and the enrichment at the capillary fringe were documented.

Figure 24 contains the results for two media whose

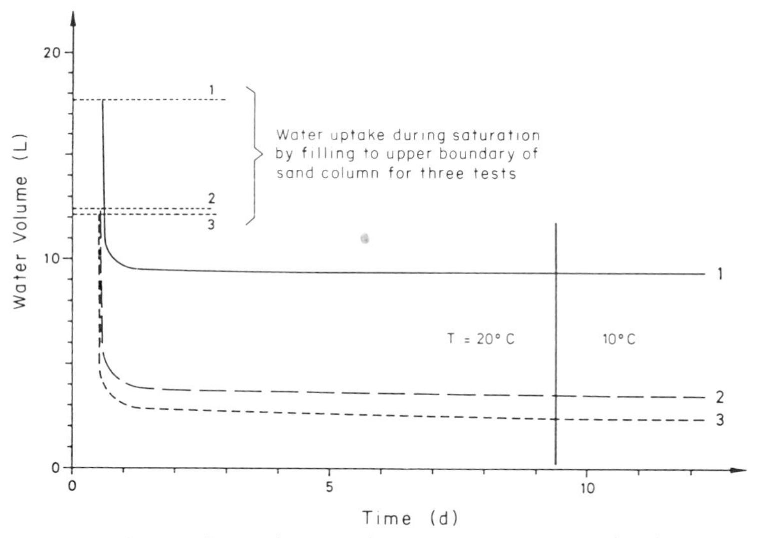

Figure 22. The outflow of water from a water-saturated column as a function of dewatering time. Column height = 117.5 cm. After 9 days, the temperature was lowered from 20°C to 10°C.

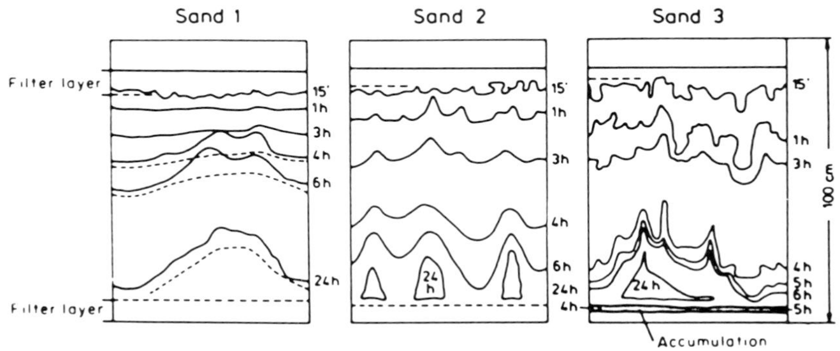

Figure 23. The progress of the perchloroethylene (tetrachloroethylene, or PER) phase front as a function of time.

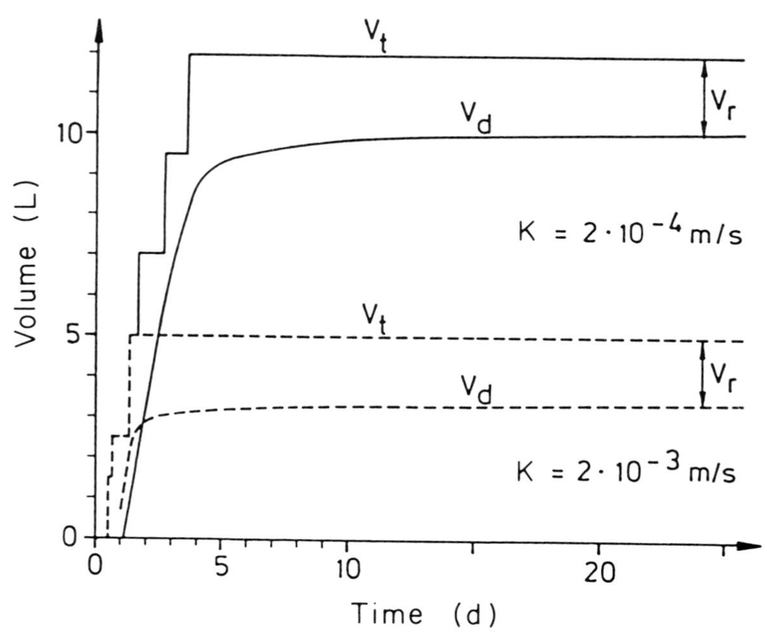

Figure 24. Determination of the residual saturation of PER as a function of the drainage time. V_t = cumulative infiltrated inflow volume; V_d = cumulative outflow volume; V_r = total retained volume (= $V_t - V_d$).

hydraulic conductivity K differed by a factor of ten. The results were obtained by the single column method (Figure 21a) with columns containing sands at their water field capacities. Both the inflow and the outflow results have been presented vs time. For both sands, the "bleed out" phase ended very quickly; the outflow curves flatten after only a few days and apparently approach an asymptotic value. The use of a maximum drain time of 20 days seems justified for such conditions.

In Figure 25, the outflow volumes for experiments with a single column with a joint (Figure 21b) have been plotted directly in terms of the retention capacity (L/m^3). A total

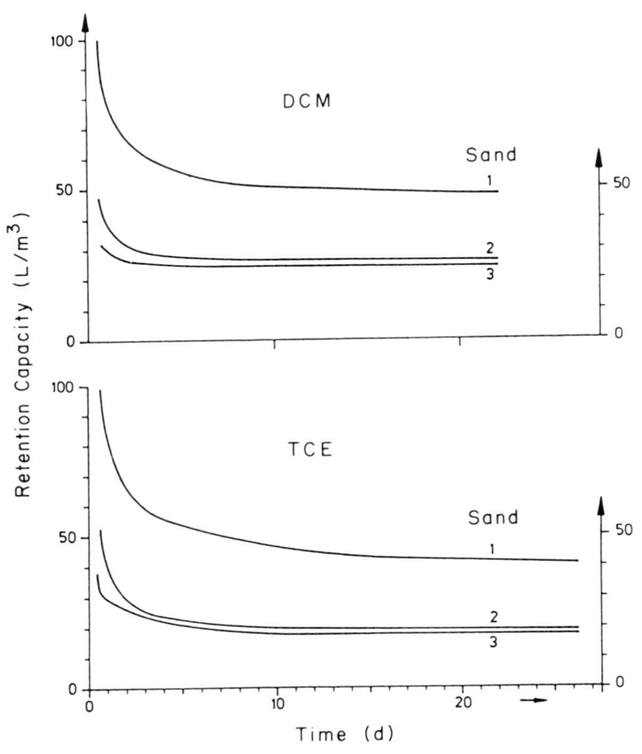

Figure 25. The retention capacity as a function of the draining time.

Table 2. Lower Bound Estimates for Residual Saturation

	Residual Saturation	
K (m/sec)	unsaturated zone (L/m^3)	saturated zone (L/m^3)
1×10^{-4}	30	50
1×10^{-3}	12	20
1×10^{-2}	3	5

draining time of 20 days was also considered adequate in these experiments.

The above procedures were modified somewhat for the determination of the residual saturation in the saturated zone. At the beginning of an experiment, the column was completely saturated (with water) up to, but not including, the upper coarse filter layer. This layer prevented the disturbance of the study medium when the CHC was applied. Its coarse grain size allowed little capillarity. The application of the CHC proceeded without the injection needle plate arrangement. The number of experiments that have been carried out with this procedure is not large; the procedure has not been worked out completely.

The results of the experiments that we have carried out to date together with the mass balance results obtained in the model trough experiments lead to the lower bound estimates for the residual saturation shown in Table 2. Due to heterogeneities, the retention capacities of unsaturated and saturated zones can, in general, be substantially higher than the tabulated numbers. This will particularly be the case for the unsaturated zone, since its water content can vary widely. It will therefore be necessary to use experience obtained with actual spills to determine what additions to the lower bound residual saturation values must be made in order to allow the reliable estimation of the overall retention capacity.

5

Microscopic Examinations

5.1 METHODS

Up to this point, we have only considered and described the behaviors of fluid chlorinated hydrocarbon (CHC) phases in terms of relatively large model systems. Such systems do not allow a testing of the extent to which the magnitudes and natures of the processes that occur around a few individual particles and their adjacent pore spaces are consistent with the data obtained for the bulk properties. The fact that microscopic examinations of the behavior of fluids in porous media can be useful has been clear since about 20 years ago; it was at that time that we were forced to start dealing with the behavior of petroleum in porous media. The experiments that we carried out at that time were, however, unsatisfactory in several respects. As a result, we sought to develop better techniques for use in experiments with CHCs.

The investigations were carried out with a macroscope (Wild and Assoc., Heerbruck, West Germany) equipped with a built-in automatic camera. While a microscope is useful for looking at small fields, a macroscope can examine relatively large surfaces with adequate magnification and sharpness. A specially prepared "model trough" served as the experimental cell (Figure 26). Three glass bars were glued to the rear glass face: one bar on the lower edge, and one each on the two side edges. These bars served as the bottom and side walls of the cell. Glass beads were used to

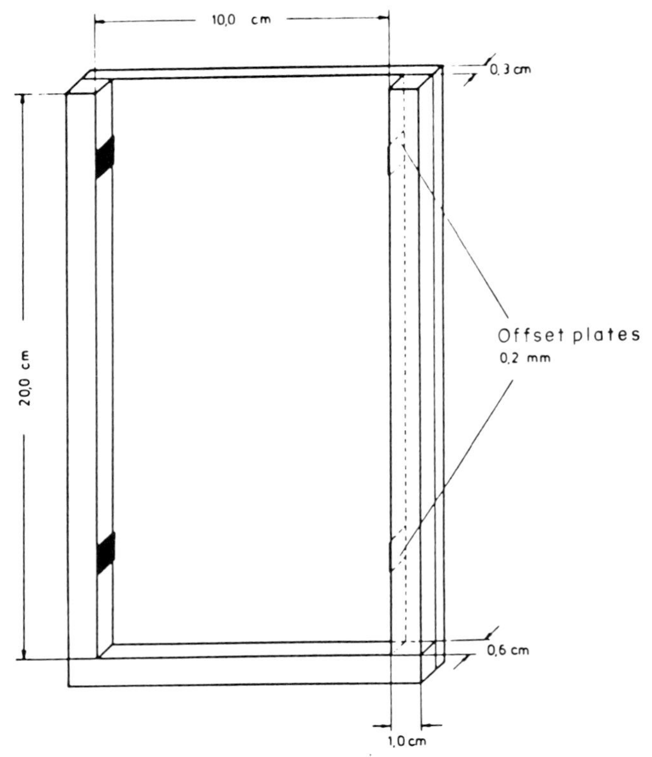

Figure 26. Frame cell for examinations with a macroscope.

form the model porous medium. With the cell at a slight angle, a single layer of beads was placed in the cell. The beads were then covered with the front plate. The latter fit into the inner dimensions of the side and bottom bars. A set of four standoff plates 0.2 mm thick were used to keep slits open between the edges of the front plate and the adjacent bars. Once filled, the cell was positioned vertically in a slightly larger glass trough. The water content of the porous medium in the cell could be regulated as desired by varying the water level in the outer trough. A jack under the outer trough was used to raise or lower the experimental setup. In this manner, the field examined by the macroscope could be moved up and down.

Soda-lime glass beads were primarily used for the porous medium. Two size ranges were used, 0.49 to 0.70, and 0.85 to 1.23 mm in diameter. Precision-graded quartz beads with similar diameters were also used but to a lesser extent. Perchloroethylene (tetrachloroethylene, or PER) was selected as the representative CHC since it is the slowest to evaporate of all the commonly used solvents.

Because the visual contrasts between air-dry beads and moist beads, and also between water and PER are small in white light, it was necessary to dye both of the fluids. For PER, the organic dye oil red at a concentration of 1 g/L was found to be the most suitable. For water, a mixture of the dyes sulfan blue and methylene blue was the most suitable. Oil red is insoluble in water, and the two water dyes are insoluble in PER.

The porous medium that was created was a vertically oriented, three-dimensional element with a thickness of only one grain. The walls of the element were the smooth walls of the front and back plates. The "sand grains" of the medium were spherical. The objection that this system provides an unrealistic model of the real world is justified to a certain degree. However, as the photos demonstrate, the great advantage of this system is that it, in fact, focuses on just one layer of grains of almost identical size and ideal spherical shape. It is only in this manner that the disruptive effects of dissimilarities can be eliminated, and textbook-clear photographs can be obtained.

A few comments are required to facilitate the understanding of the photos (Figures XIIIa-XXb). The glass beads appear as circles. A substantial amount of magnification was used. The photos were usually focused on either the front side of the beads (i.e., the side touching the front plate of the cell), or on the plane passing through the centers of the beads. Focusing on the rear sides of the beads was only possible for beads that were transparent.

It should be noted that when water (blue) and air alone are

present, water is the wetting phase. When a CHC (red) and air alone are present, the CHC is the wetting phase. When all three are present, water is the wetting phase, and the CHC is the nonwetting phase. When some residual air is present in the system, both water (Figure 27a) and PER (Figure 27b) will accumulate in the wedge-like spaces at the bead/bead contact points and at the bead/glass plate contact points. When viewing a given system, although the same capillary effects are operating, one obtains different perspec-

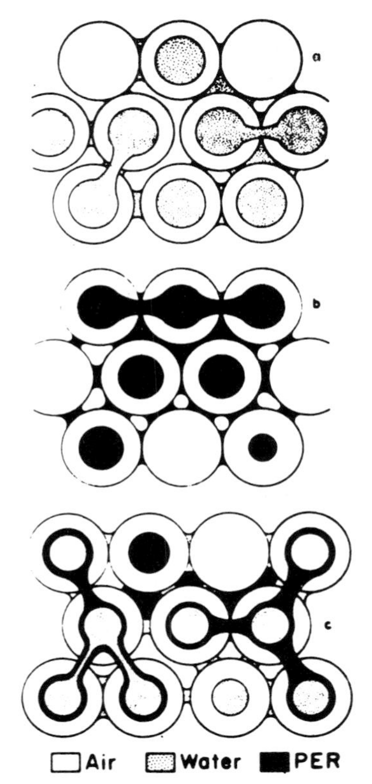

Figure 27. Fluid distributions in the unsaturated zone (schematic). Bridge formation is at the bead/bead and bead/plate contact points. a) Bridge formation by water. b) Bridge formation by PER. c) Bridge formation by water together with PER.

tives when viewing the system differently. When viewing the bead/bead contact points, one observes the rings of fluid at the contact points from the side. When viewing the bead/plate contact points, the perspective is normal, and the rings of fluid appear circular. The circular accumulations of fluid can merge into one another, thereby forming striking geometric patterns. At the bead/plate contact points, accumulations of fluid form that are halves of what would be present at bead/bead contact points. The two half-accumulations on either side of each bead are mirror images of one another and parallel to the plates.

In moist media, water wets the surfaces of the beads, and if enough water is present, also occupies the wedges of pore space around the beads. CHC that then penetrates into the system must make do with whatever void pore space remains. In dry media, PER will likewise first take firm possession of the wedges around the beads. When water then infiltrates into the system, it will begin to (1) displace the CHC masses down over the beads; and (2) drive the bead-to-bead bridges of CHC into droplets that will occupy the wedges of pore volume between the beads. The CHC that is wetting the surfaces of the beads will thereby be displaced, and the beads rewetted with water. The overall rewetting/displacement process occurs relatively quickly. It is presented schematically in Figure 28. The droplets forced into the pore wedges may either remain there or move downward along pathways wetted with CHC.

5.2 EXAMPLES

Figure XIIIa

The medium was initially dry. The bead diameter range was 0.85 to 1.23 mm. Water was then dripped in from above. The pore corners are seen to be partly filled with water, recognizable by its deep blue color. The smaller pores in the

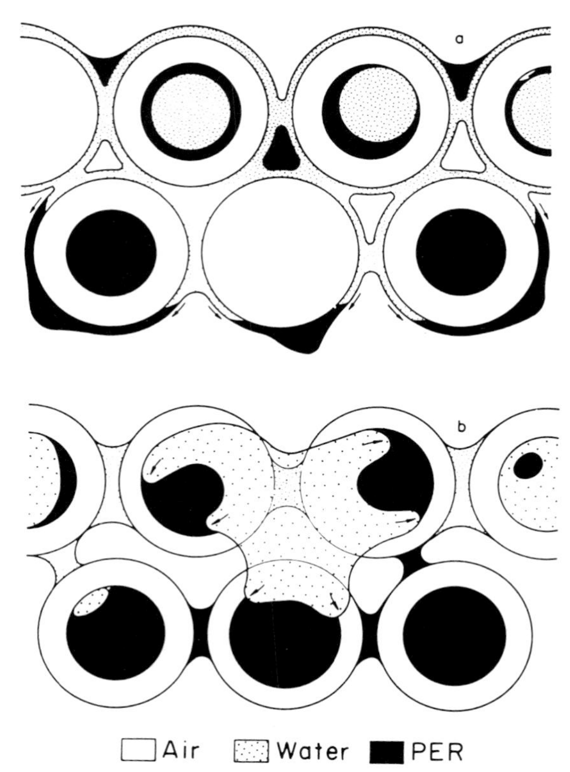

Figure 28. Displacement of PER in the unsaturated zone by water (schematic). a) Detachment of PER from the beads as viewed in a plane through the bead centers and parallel to the plates. b) Displacement of PER on the bead hemispheres in the neighborhood of the front glass plate.

right half of the picture are filled with water. All other pores contain air.

Figure XIIIb

The medium was initially dry. The bead diameter range is 0.85 to 1.23 mm. As in Figure XIIIa, water was dripped in from above. The disk-shaped volumes of fluid at the bead/

plate contact points have, in part, joined together. The smaller pores contain water while the remaining pores still hold air.

Figure XIVa

The medium was initially dry. The bead diameter range was 0.85 to 1.23 mm. PER was dripped in from above. Most of the pore corners contain PER. The plane of bead/plate contact points is clearly distinguishable. The remainder of the pore volume contain air.

Figure XIVb

The medium was initially dry. The bead diameter range is 0.85 to 1.23 mm. PER was dripped in from above. Some of the disk-shaped volumes of fluid at the bead/plate contact points became joined. The small clear drops of PER in the upper left corner were formed by condensation of PER vapor.

Figure XVa

The medium was initially moist. The bead diameter range was 0.85 to 1.23 mm. PER was then dripped in from above. The pore corners were already occupied by water. As soon as the supply of PER was discontinued, the threads of PER became detached and quickly ran down through the medium. The PER remaining is present in the form of drops that conform to the dimensions of the pore spaces. It remains hanging in the smaller pore spaces in a state of residual saturation, and as such, is immobile in the unsaturated zone.

Figure XVb

The medium was initially dry. The bead diameter range was 0.85 to 1.23 mm. PER was added and occupied the corners of the pores. The medium was then saturated from below with water. The water forced the PER into the smaller pore spaces. The drops of PER have taken on a spherical-like shape there and conformed in part to the available pore spaces. At this point, the PER is in a state of residual saturation in the saturated pore space.

Figure XVIa

The medium was initially moist. The bead diameter range was 0.85 to 1.23 mm. PER was dripped in from above. It has accumulated as a thin sheath around a zone of higher water content. An enrichment of PER is observable in the figure in the "bays" of the sheath.

Figure XVIb

The medium was first saturated with water. The bead diameter range was 0.49 to 0.70 mm. PER was applied from above. The PER penetrated in a fingerlike stream into the saturated zone. This stream held together so long as the flow of PER continued. When the supply of PER was discontinued, the front portion of the finger was torn from the stream, flowed downward for a short period of time, then came to a standstill. The figure illustrates the PER in an isolated and immobile state. This type of condition will comprise a state of residual saturation, but it cannot be excluded that such a mass of CHC will not break up into smaller pieces.

Figure XVIIa

The medium was initially dry. The bead diameter range was 0.85 to 1.23 mm. When PER was added, it occupied the corners of the pores and was also present as films on the bead surfaces. Water was then applied from above, but the medium did not become saturated with water. In the figure, the infiltrating water is seen to be driving the PER films (red) together. Lower right: the detachment of a PER film from a bead begins. Upper right: the water has already detached the PER from the bead and begins to surround the displaced PER.

Figure XVIIb

The medium was initially dry. The bead diameter range was 0.85 to 1.23 mm. When PER was added, it occupied the corners of the pores and was also present as films on the bead surfaces. Water was then applied from above. The films of PER are seen to have been displaced by the water and forced to occupy rings around the contact points. The processes depicted in Figures XVIIa and XVIIb represent the replacement of the nonwetting fluid by the wetting fluid. (In the presence of air, a CHC will remain the wetting fluid.)

Figure XVIIIa

The medium was initially saturated with water. The bead diameter range was 0.85 to 1.23 mm. PER was applied at the upper boundary of the capillary fringe. The figure illustrates how it spread out there into a thin layer.

Figure XVIIIb

The medium was initially saturated with water. The bead diameter range was 0.85 to 1.23 mm. More PER was added to that placed at the top of the capillary fringe in Figure XVIIIa. As a result, the PER is now seen to have penetrated deeper into the medium.

Figure XIXa

A two-layer medium was initially saturated with water. The bead diameter range in the upper layer was 0.49 to 0.70 mm. In the lower layer, it was 0.85 to 1.23 mm. PER was applied from above. In the layer with the smaller pores, the infiltration paths were broader than in the layer with the larger pores. After the supply of PER was discontinued, the PER streams became disconnected at the interface between the two layers. This led to a stranding of some PER in a state of residual saturation.

Figure XIXb

A two-layer medium was initially saturated with water. The bead diameter range in the upper layer was 0.85 to 1.23 mm. In the lower layer, it was 0.49 to 0.70 mm. PER was applied from above and sank in the upper layer (large pores) on infiltration paths that were relatively narrow. The latter are not visible in the figure. When the PER reached the fine-pored layer, it spread itself out laterally and became dammed up. Only after a certain PER pressure built up did it begin to slowly penetrate into the small-pored layer.

Figure XXa

The medium was initially saturated with water. The bead diameter range was 0.85 to 1.23 mm. The PER moved around an "island of water" for no readily apparent reason. Such random pathways are also visible in the lower third of Figure XXb.

Figure XXb

The medium was initially saturated with water. The bead diameter range was 0.49 to 0.70 mm. The PER sank in a completely random manner. While it certainly chose the pathway of least resistance, that pathway cannot be predicted even by an observer with a practiced eye. When the PER supply was discontinued, the PER stream remained intact and did not break up into smaller residual masses. Figures XXa, XXb, and 29 serve as examples for the unpredictability of the paths of CHCs in the subsurface.

5.3 INTERPRETATION

The instructive photos discussed above illustrate the behavior of CHCs on the microscale. It is now useful to briefly discuss the principle of the simultaneous movement of two immiscible fluids in a porous medium. This phenomenon is called *two-phase flow,* and it is often not understood correctly. Figure 30 is a diagram that gives the relative permeability of a porous medium to two immiscible fluids as a function of the degree of saturation of the fluids. This type of diagram is well known in the petroleum recovery field. For each fluid, the permeability of the medium to that fluid is referenced to the permeability for the pure fluid. It should be noted that the relationships between these relative permeabilities and the degrees of saturation are by no means linear. Also, the curves for the two fluids do not begin to rise above zero permeability at $S = 0$ and $S = 100\%$, respectively. Rather, they increase from zero at S_{01} and S_{02}. The two fluids hinder each other's movement to different degrees. Both fluids must reach a minimum saturation before they achieve any mobility at all. Water will not flow until its degree of saturation reaches at least S_{01}. Prior to reaching S_{01}, it is in a state of *irreducible saturation*. The nonwetting CHC will not flow until S_{02} is reached. Prior to reaching S_{02}, the CHC is in a state of residual saturation and is not mobile under normal pressure conditions. The curves also demonstrate that at S_{02}, the mobility of the water is much below 100% of its pure state value. At S_{01}, on the other hand, the mobility of the

Figure 29. Sinking of PER in a saturated medium. The photo was made with a macroscope. The bead diameter range was 0.49 to 0.70 mm. The PER pathway proceeded randomly. (See also Figure XXb.)

CHC is not decreased substantially below its pure state value.

This means that water levels in the irreducible range do little to hinder the flow of CHC. In contrast, when a CHC is present at residual saturation values, the flow of water is restricted substantially. Without microscopic observations of a porous medium containing CHC and water, it would be very difficult to understand the effect of the CHC on the water permeability. It would be necessary to derive it theo-

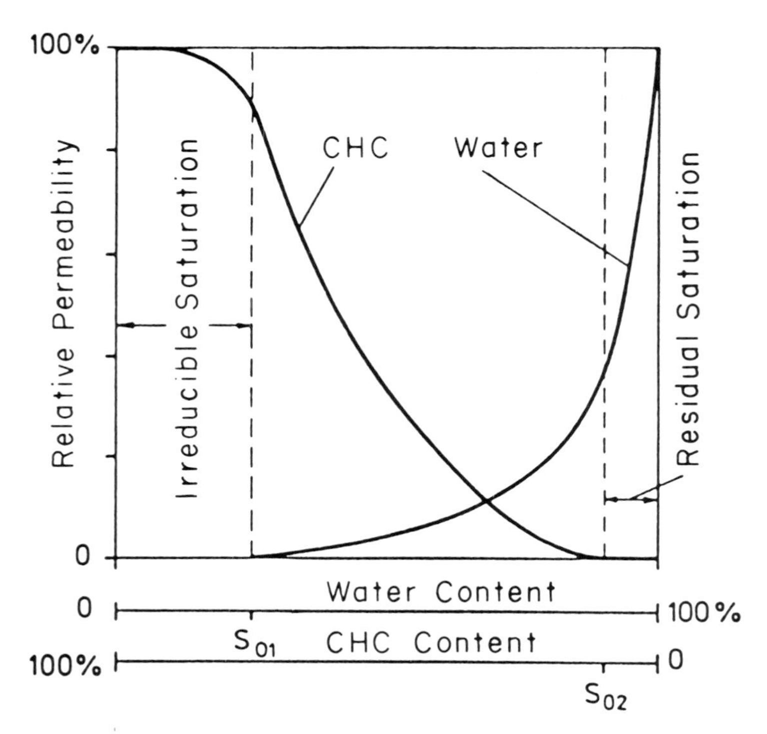

Figure 30. The relative permeability curves for water and a typical CHC in a porous medium as a function of the pore space saturation.

retically. The microscopic observations carried out here illustrate that when a nonwetting fluid is introduced into a moist medium, a CHC will not be able to occupy the corners of the pores; they will already be occupied by water. The CHC will be forced to remain in the open portion of the pore volume where it will hinder water flow. In a dry medium, the CHC will occupy the corners of the pores. If water is then added, however, the CHC will again be forced into the open portion of the pore volume.

Figure 30 can only indicate the general trends according to which the permeabilities of the two fluids will vary as a function of the composition of the pore space. The same statement applies to the situation when gas (air) is present as a third phase, in which case a triangular permeability diagram must be drawn. These issues will not be considered here further, other than to reiterate that with multiphase flow, the permeability of a medium to a fluid will be highest when that fluid occupies 100% of the pore space. It is thus impossible to predict the relative flow velocities of the fluids based simply on their kinematic viscosities.

In the case of multiphase flow involving petroleum as one of the phases, the relationships between the relative permeabilities and the degree of saturation have been investigated in thousands of experiments. To our knowledge, the nature of these relationships have not been verified experimentally for CHCs as nonwetting phases. The diagram in Figure 30 has therefore been adapted from results obtained with petroleum materials for use for the first time with CHCs.

6

Spreading as a Fluid Phase in a Fractured Medium

6.1 METHODS

In approximately half of West Germany, bedrock lies close to the ground surface. In these rocks, the groundwater flows preferentially or at least partially through fractures, or in actual channels as in karst formations. For the purposes of this book, the term *fractured media* will be used to refer collectively to all media that contain fissures and joints. It will include both natural fractured rocks as well as artificial fractured materials such as fractured concrete. We felt it was very important to undertake the investigation of the behavior of chlorinated hydrocarbons (CHCs) in fractures with model experiments since, to our knowledge, this has not yet been done elsewhere.

Fundamentally, the same physicochemical laws apply to the behavior of CHCs in rock fractures as in nonindurated media. One marked difference between the two types of media, however, is that the finely branched network of capillaries in a porous medium cannot be compared to a network of fractures with approximately parallel walls. In addition, in wide fractures and in the channels of karstic media, a different type of flow can occur.

The examination of the spreading processes of a CHC in a fractured medium can be carried out in a single fracture of adequate dimension. (To a large extent, it is not necessary to set up a complete, technically complex, cross-connected fracture network.) For each experiment, two 1-m^2 plates of construction grade glass were used. One of the plates was firmly mounted to a base plate that was resistant to CHCs. The second plate was mounted to the first plate with screw clamps. The desired fracture aperture c was set with thin "off-set" plates held between the two glass plates. The bottom gasket consisted of a metal strip of the same thickness as the off-set plates. The bowing out of the pair of plates (or even their breaking) when filled was avoided by placing the apparatus in a glass-walled trough the plates of which could withstand the pressure of a 100-cm-high water column. The internal and external pressures on the plates of the fracture apparatus were therefore equal when the trough was full. The spreading of the fluids in the model fracture were studied with a time-lapse camera. This was carried out by photographing the model system directly through the water in the trough. The effects of hydraulic roughness of the fracture surface was studied by using glass sheets that were: (1) unaltered in their surface properties (hydraulically smooth); and (2) roughened by sandblasting (hydraulically rough). The effects were studied in preliminary experiments carried out with an apparatus with a 70-cm-high fracture and both smooth and rough walls.

The manner in which infiltration occurred in an experiment with perchloroethylene (tetrachloroethylene, or PER) ($\rho = 1.62$), fracture aperture $c = 0.2$ mm, and hydraulically smooth walls is presented in Figure 31. The unsaturated zone was penetrated quickly by a narrow core of PER. When the capillary fringe was reached, it first caused a lateral spreading of the PER. Following this, on the lower boundary of the PER that accumulated on the capillary fringe, fingers of PER formed. Once formed, these fingers penetrated the saturated zone quickly all the way to the bottom of the system. At that point, they then joined to form a low mound of PER. This mound slowly spread itself laterally in a manner similar to that observed in the large model trough (Figures 6 and 7). At the end of the infiltration, only small residuals of PER remained in the unsaturated zone. Other than the

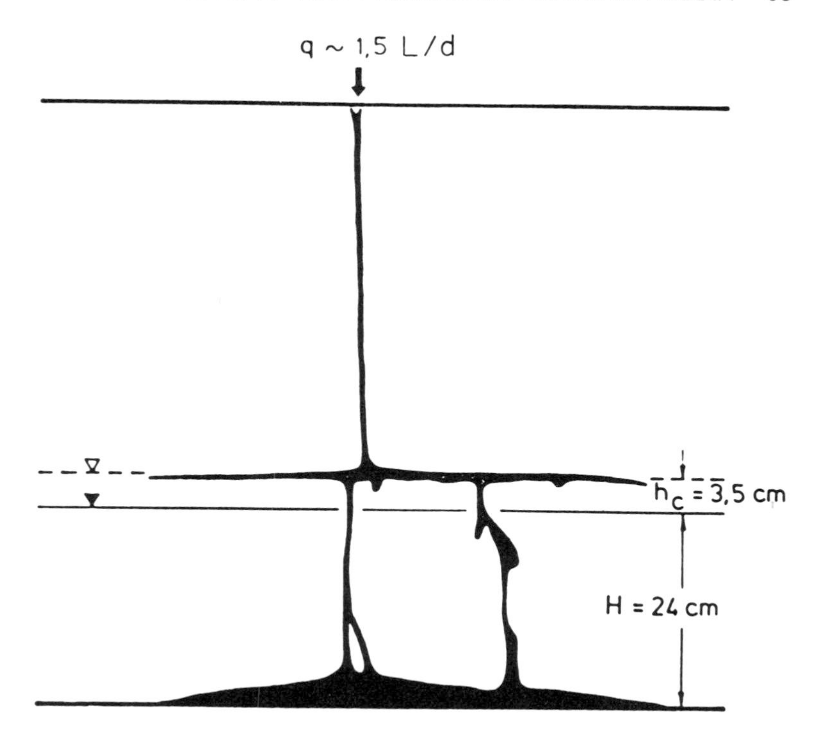

Figure 31. Fracture model experiment with PER during infiltration. Hydraulically smooth condition and c = 0.2 mm.

mound of PER accumulated on the bottom of the system, the same was true for the saturated system.

The manner in which infiltration occurred with PER, fracture aperture c = 0.2 mm, and hydraulically rough walls is presented in Figure 32. In contrast to the hydraulically smooth case, the rough surfaces caused a broad and very finely veined flow net. At the end of the infiltration, numerous droplets and films remained in the unsaturated zone. The same was true for the saturated zone, but, in addition, some larger flecks of CHC also remained.

In the hydraulically smooth fracture (Figure 31), the PER penetrated in narrow cores and in a surprisingly quick manner; a fracture aperture larger than 0.2 mm was therefore not

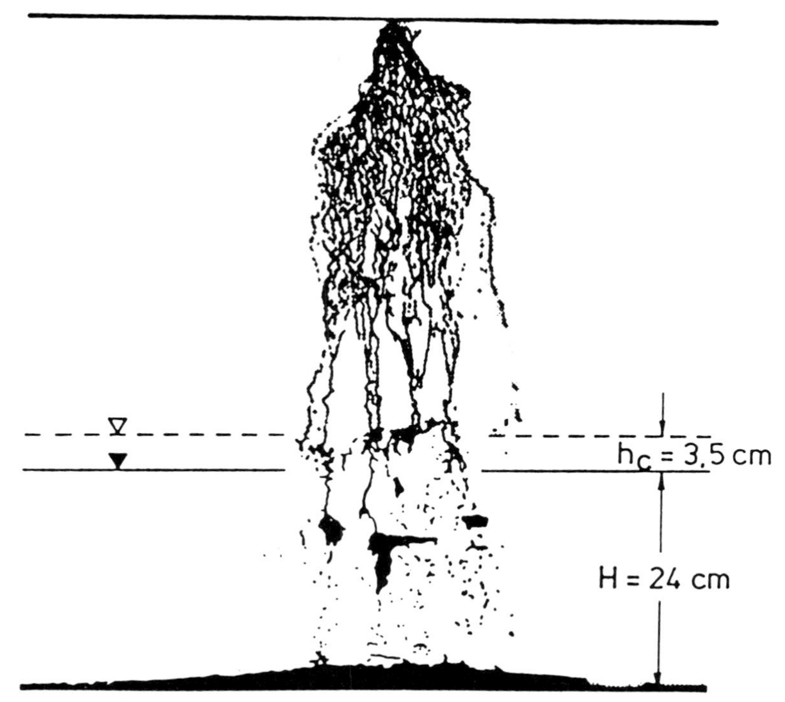

Figure 32. Fracture model experiment with PER during infiltration. Hydraulically rough condition and c = 0.2 mm.

of interest in the subsequent experiments. At this point, c was decreased to 0.1 mm (Figure 33). The PER infiltration rate was also cut back by about one-half in order to allow for the decreased capacity of the smaller fracture. During the infiltration, a relatively broad body of CHC was initially formed. Nevertheless, at the lower boundary of that body, several fingers of CHC formed and quickly extended themselves downward. As soon as they reached the upper boundary of the capillary fringe, lateral spreading on the capillary fringe was initiated. Soon thereafter, PER reached the edges of the pair of glass sheets and started to drip out into the water in the model trough. The whole spreading

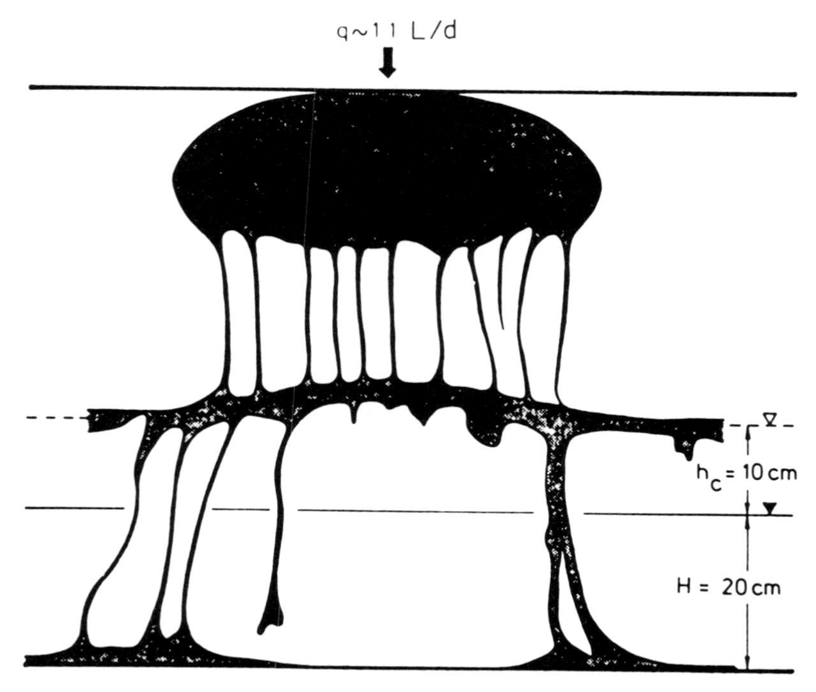

Figure 33. Fracture model experiment with PER during infiltration. Hydraulically smooth condition and c = 0.1 mm.

was accompanied by only a minimal amount of depression of the water in the capillary fringe.

As the lateral spreading of the PER layer took place on the capillary fringe, fingers formed on the bottom of that layer. These fingers then proceeded to penetrate the saturated zone. They collected and spread on the bottom of the system. At the end of the infiltration of the PER, the jellyfishlike PER body in the unsaturated zone bled itself out quickly; only a thin film of PER was left behind on the walls. Subsequently, the PER layer on the capillary fringe ruptured, leaving behind several large fragments of PER. The fingers in the saturated zone were then finally depleted; numerous small flecks of PER marked the locations of their paths.

In these experiments, the fracture aperture values were set

carefully since this parameter affects the retention capacity of fractured systems. With hydraulically smooth walls, reasonable retention capacities for CHC can occur with fracture apertures smaller than 0.2 mm. The experiments discussed in Section 6.2 were carried out for both smooth and rough walls but only with the apertures 0.2 and 0.1 mm. Values of c smaller than 0.1 mm would not have been obtainable without a substantially greater degree of technical effort.

Unfortunately, the experiments that are described above and in Section 6.2 suffered from a deficiency; strictly speaking, the information obtained can be used only in situations involving fractures in sandstone and quartzite. Nevertheless, a transfer of the results to situations involving fractures in limestone and dolomite will probably also be permissible. We are not able to make any comments about the transferability of the results to cases involving fractures in shale or marl, or fractures the walls of which consist of clay or ferriferous materials. Therefore, we are considering the possibility of conducting wetting experiments with test specimens of actual fractured materials.

6.2 EXAMPLES

Some typical photos of the spreading of PER in a model fracture are presented in Figures XXIa–XXIVb. The fracture surface was 100 × 100 cm. The CHCs were dyed red with oil red at a concentration of 1 g/L.

Figure XXIa

PER, c = 0.2 mm, hydraulically rough, H = 93 cm, h_c = 5 cm, T = 17° C. The fracture was saturated with water almost to the top of the fracture (H + h_c = 98 cm). The PER was applied at a single point directly on top of the capillary fringe at a rate of 1.4 mL/min. PER sank within 2 min through the saturated zone through a narrow region, and showed only a small amount of dendritification. At the bottom of the system, it formed a low mound (maximum height = 3 cm) and proceeded to escape into the outer trough. The PER was applied incrementally. After each increment, a strand of PER passed down through the fracture and separated itself from the application point as well as from most points along the walls. This process served to temporarily leave behind isolated flecks of PER on the walls. When these flecks grew to larger sizes and weights, they became mobile again and penetrated to the bottom along the preferred path. After 11 min, the application of the PER was suspended since the PER distribution was no longer changing. A short period of time thereafter, only a few flecks of PER with sizes up to about a centimeter remained. Therefore, the retention capacity of PER in a saturated fracture with c = 0.2 mm will be very small.

It is also interesting to discuss what occurred when the water level was lowered in the fracture. As soon as the upper boundary of the capillary fringe (which was also the upper boundary of the saturated zone) reached a fleck of PER, the PER fleck was forced from the saturated zone into the unsaturated zone. At this point, it separated from the walls and sank into the saturated zone. In that region, it was able to combine with other flecks, and if a resultant aggregate fleck became large enough, it would in turn sink down further. At the conclusion of this process, the fracture was left unsaturated, and only a negligible amount of PER remained in the fracture.

This and other experiments carried out to study these effects indicate that lowering the water level in a fracture causes a CHC that has become immobilized as flecks on the fracture walls to become partially remobilized. In this manner, the unsaturated fracture space (c = 0.1 to 0.3 mm) will become practically free of CHC. To the extent that wetted paths of CHC exist on the underlying saturated (water) walls due to the previous infiltration, the mobilized CHC will sink along those paths. The lowering of the water level by means of wells after a real spill will therefore be of little use. Indeed, it may prove disadvantageous since it will cause a downward transport of CHC.

After the water level was lowered to a point such that H = 24 cm, the thickness of the unsaturated zone was 71 cm. At this point, PER was applied again at the previous infiltration rate. The PER sank along two or three narrow paths at a velocity larger than that observed earlier. It was dammed up at the capillary fringe only briefly and moved rapidly to the bottom of the system. At the end of the application, a small number of PER droplets remained only in the saturated zone.

Figure XXIb

PER, c = 0.1 mm, hydraulically rough, H = 94 cm, $h_c > 6$ cm, T = 14°C. The fracture was saturated with water all the way to the top. The PER was applied to the top of the capillary fringe at a rate of 1.2 mL/min. This rate was slightly slower than that used in the comparable experiment with c = 0.2 mm (Figure XXIa). The path down which the PER moved was noticeably broader than in the previous experiment (Figure XXIa). In view of the small fracture aperture, however, the path width was still relatively narrow. The bottom of the system was reached in only 6 min. The application of PER was terminated after 25 min because at that point the distribution of PER was no longer changing. At that time, numerous flecks of PER remained along the path the PER had followed to the bottom. The total amount of PER that remained was estimated roughly to be a few milliliters.

Figure XXIIa

PER, c = 0.1 mm, hydraulically rough, H = 23 cm, $h_c = 12$ cm, T = 17.5°C. The fracture was first saturated with water all the way to the top. The water level was then lowered to 23 cm above the bottom of the system. The application of the

PER started 1 hr after the water level was lowered; a relatively large number of water flecks wetting the fracture walls still remained at that point. The PER application rate was 1.3 mL/min. The water flecks affected the distribution of the PER above the water table.

In the unsaturated zone, the PER path broadened itself quickly and somewhat asymmetrically as it moved downward. The asymmetry was possibly due to a nonuniform distribution of residual water or to a nonuniform fracture aperture. The infiltration front was not smooth. Rather, the PER moved downward along numerous finely divided veins. At the capillary fringe, the PER dammed up and spread itself laterally. At the lower boundary of the accumulated PER, however, protuberances were formed rapidly. A few of these protuberances grew in size and penetrated rapidly to the bottom of the system. It is worth noting that while a comparatively closely meshed flow net of PER pathways formed in the unsaturated zone, only a few (and therefore larger) fingers of PER developed in the saturated zone.

Figure XXIIb

(Continuation of experiment in Figure XXIIa.) This photo illustrates the conditions present 1.5 hr after the discontinuation of the infiltration of the PER. At this point, the PER in the fracture was immobile; the PER distribution was no longer changing in time. This experiment demonstrated that the retention capacity of a 0.1-mm fracture will not be significant in either the unsaturated or saturated zones. For the portion of the unsaturated zone that was wetted with PER, a rough mass balance led to a residual saturation of 0.05 L/m^2 of fracture surface. Therefore, 1000 m^2 of fracture surface can retain 50 L of PER. This is very small in comparison to what can occur in a porous medium. For example, this amount could be held in residual saturation in 1 m^3 of fine sand with a hydraulic conductivity of 1×10^{-4} m/sec. The 0.05 L/m^2 value cited above should only be used to provide an order of magnitude appreciation of the retention capacity of narrow unsaturated fractures. A large number of additional experiments are required in order to acquire useful values that can be applied in practical problems.

Figure XXIIIa

Dichloromethane (DCM), c = 0.1 mm, hydraulically rough, H = 94 cm, h_c > 6 cm, T = 20.4°C. The DCM application rate was 1.4 mL/min. Excluding the fact that the DCM distribution at the application point was not the same as that observed in Figure XXIb, the sinking of the DCM toward the bottom of the system was otherwise quite comparable to that observed in Figure XXIb. The DCM reached the bottom of the system in only 6 min. After 8 min, it reached the left edge of the fracture, and it reached the right edge after 13 min. At those edges and points in time, the DCM began to drip out of the fracture. The addition of the DCM was discontinued 18 min after initiating the application. Approximately 30 min after ending the application, the remaining DCM was practically immobile, and residual saturation was reached in the saturated zone. It should be noted that some flecks with areas as large as 2 to 3 cm^2 were present.

Figure XXIIIb

(Continuation of experiment in Figure XXIIIa.) Approximately 45 min after ending the application of the DCM in the previous experiment, the water level was lowered to H = 24 cm. After so doing, very little DCM remained in the newly created unsaturated zone. This was due to downward movement as well as evaporation of the DCM. At this point, an additional 25 mL of DCM were added in a manner similar to that used above. The temperature (T = 21.8°C) was slightly higher than above (20.4°C, Figure XXIIIa).

After only 3 min, the DCM penetrated through the saturated zone, and a single finger of solvent reached the bottom of the system. A few minutes later, three additional fingers of solvent reached the bottom. The figure shows the conditions after 15 min. The DCM distribution in the unsaturated and saturated zones is very comparable to the PER distribution in Figure XXIIa. The experiment was stopped after 18 min.

Figures XXIVa and XXIVb

(Continued discussion of experiment in Figure XXIIIb.)

These photos show enlarged sections of the unsaturated zone (Figure XXIVa) and the saturated zone (Figure XXIVb) as they appeared approximately 10 minutes after the addition of DCM was discontinued for Figure XXIIIb. These figures again illustrate that the DCM was distributed as small flecks in the unsaturated zone and as larger flecks in the saturated zone. A nearly water-free layer of DCM was formed on the bottom of the system.

Although the scenarios discussed in Chapter 2 for the spreading of CHC will also apply in a fundamental sense to fractured media, substantial qualifications will generally be required. For example, the retention capacities of fractured rocks will vary by orders of magnitude, and the presence of only a few fractures with apertures of the order of only 0.5 mm will be able to control the infiltration of CHC into the ground. With fractured rocks, then, it will be absolutely necessary that the subsurface spreading of CHC be analyzed on a case by case basis and that all of the likely geological and tectonical conditions be taken into consideration.

On the basis of the above results, it appears unlikely that CHC will be retained in meaningful amounts in natural fractures with apertures of greater than 0.2 to 0.5 mm. This applies to both unsaturated and saturated systems. In the case of rocks containing fractures smaller than 0.3 mm, a modest retention capacity can be obtained but only when the rock is intensively fractured.

Finally, it is important to point out that for our latitudes, where fractured rocks exist in urban development and transportation line areas, those rocks are also more or less overlain by substantial thicknesses of nonindurated media. Being porous, the latter can possess substantial retention capacities. Such combinations of porous media with fractured media can result in very complicated infiltration scenarios. [Translators Note: An additional complication will result when the fractured rock is also itself somewhat porous. In this case, matrix diffusion of a dissolved CHC to that porosity can cause increased retention and, therefore, retardation relative to the velocity of the water in the fractures.]

COLOR PLATES I Through XXIV

DENSE CHLORINATED SOLVENTS IN POROUS AND FRACTURED MEDIA

Model Experiments

48 Full-Color Photographs

I.a.

Chlorinated Hydrocarbons (CHCs)

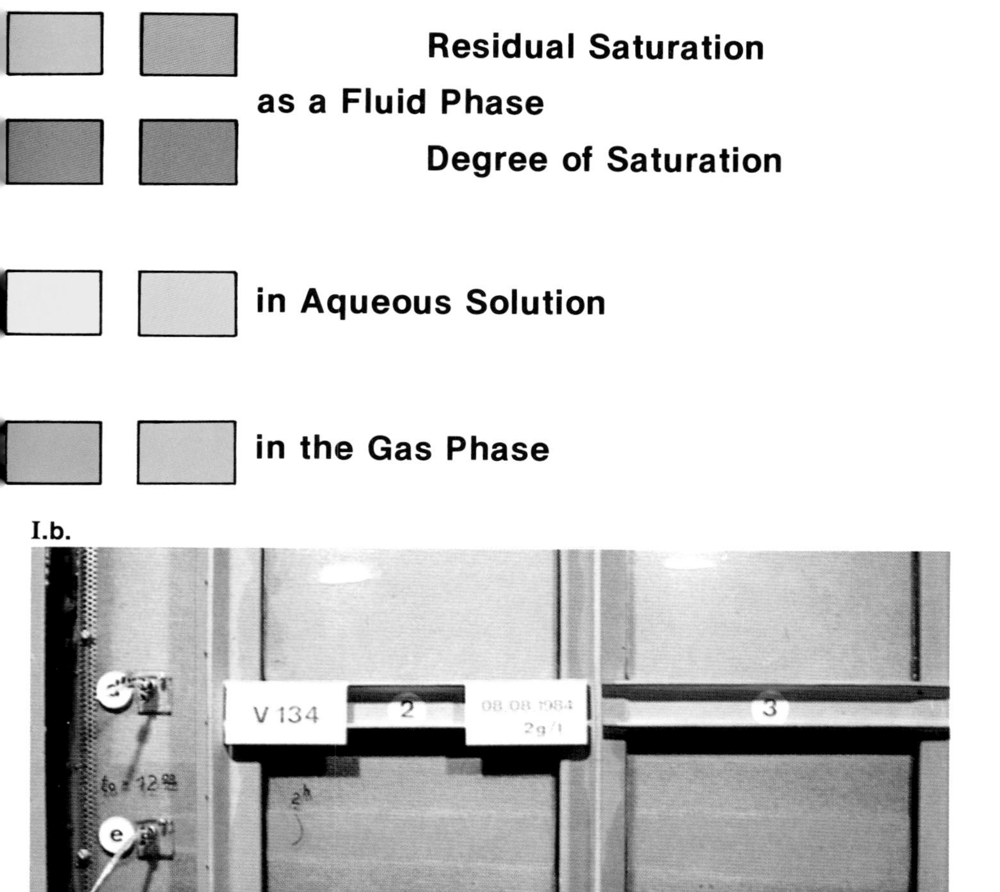

I.b.

Figure I.a. Key for phase type and location of chlorinated hydrocarbons (CHCs). Darker colors indicate higher concentrations.

Figure I.b. Streamlines of fluorescein solution in experimental trough demonstrating parallel flow lines near injection points.

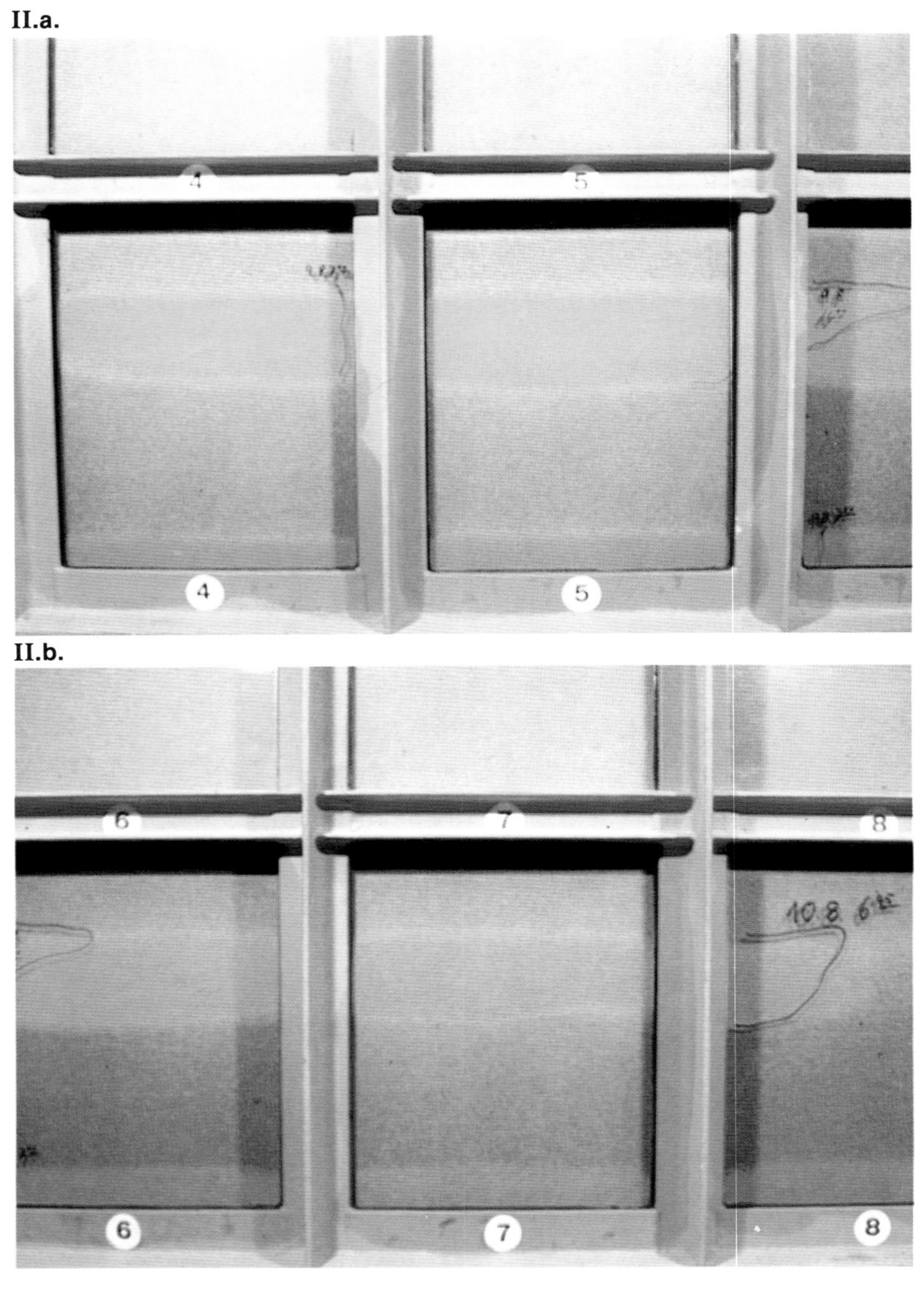

Figure II.a. Same as Figure I.b. except further down the system.

Figure II.b. Same as Figure II.a. except further down the system.

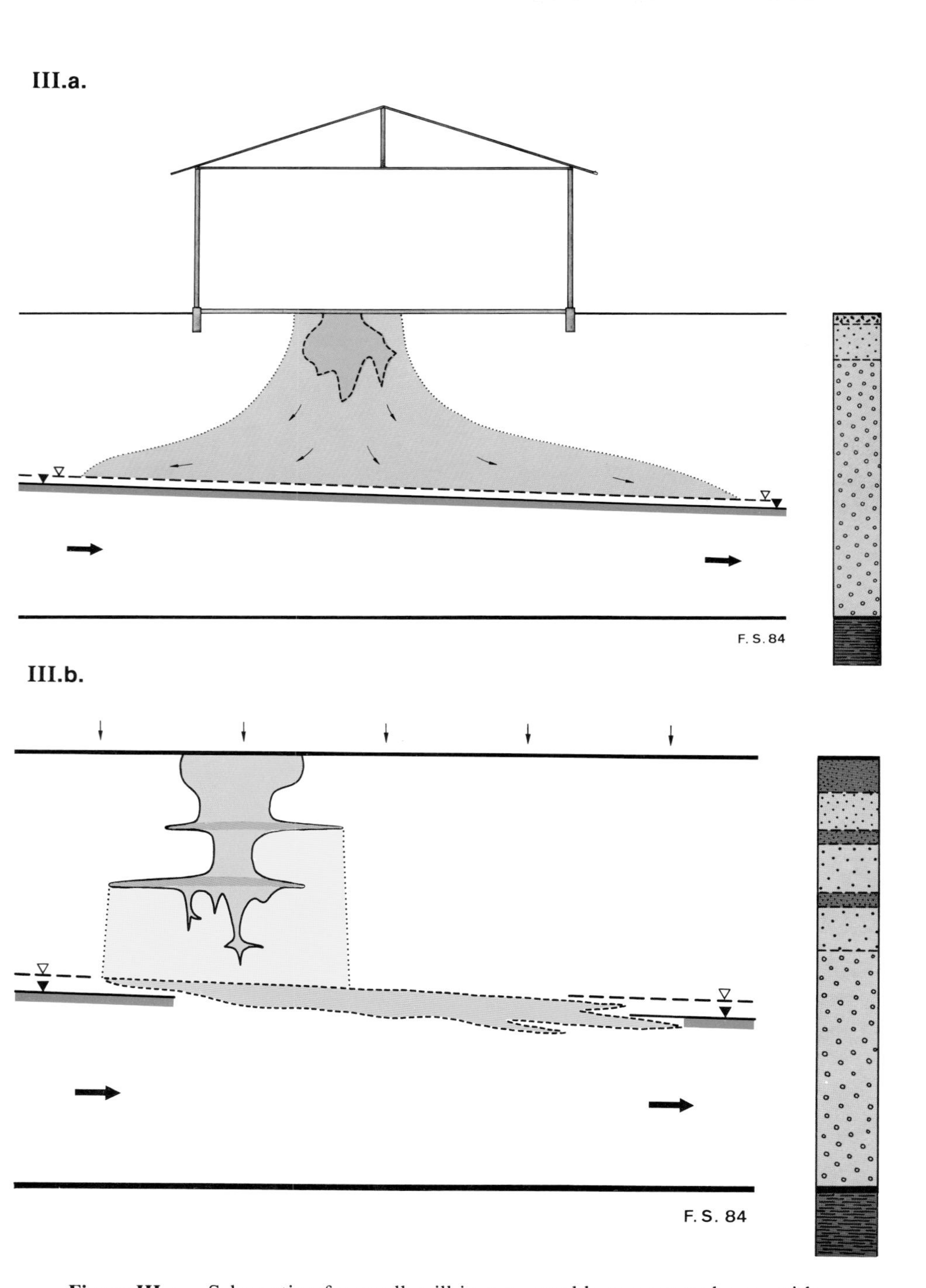

Figure III.a. Schematic of a small spill in a permeable unsaturated zone with a resulting mound of CHC gas. Concentration decreases with distance from spill. Grainsize shown at right.

Figure III.b. Larger spill than in III.a., but still not large enough to exceed the retention capacity of the unsaturated zone; no liquid CHC reaches the capillary fringe. Grainsize at right.

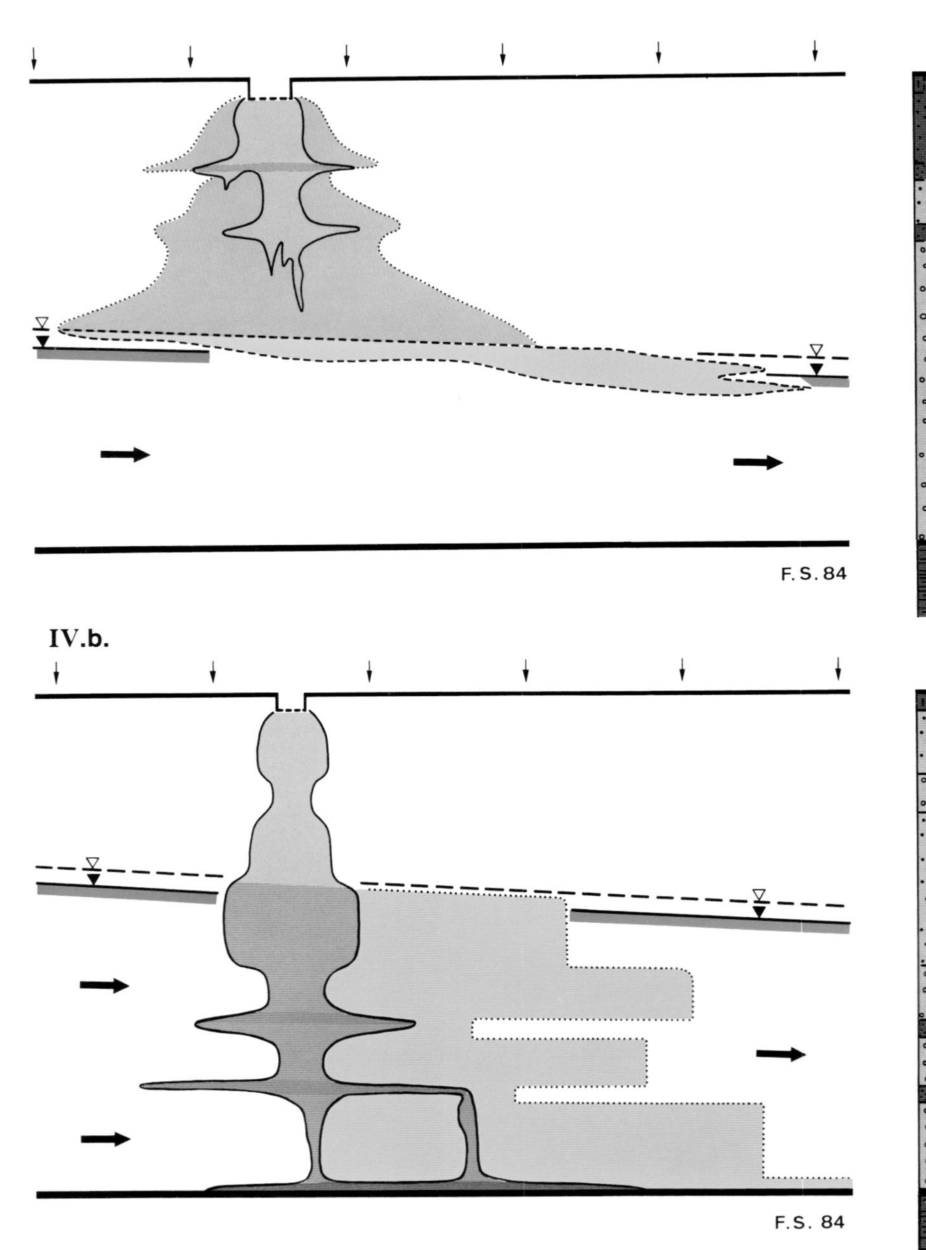

Figure IV.a. An envelope of CHC vapor develops around the core of the Figure III.b. spill, leading to a gas mound. Liquid and gaseous CHC tend to dam up on the top of low permeability layers. Grainsize at right.

Figure IV.b. Percolated mass of CHC exceeds the combined retention capacities of the unsaturated and saturated zones. Liquid CHC reaches the bottom confining bed and accumulates there. Groundwater velocity depends on local permeability.

V.a.

V.b.

Figure V.a.
Three 1-m-high columns used to determine infiltration velocities and retention capacities in porous media of different permeabilities.

Figure V.b.
Column with dyed PER in a sand filter zone. Infiltration into test sand underlying filter zone is prevented by the fine grain size of the test sand.

VI.a.

VI.b.

Figure VI.a.
Entry of PER into a fine grain size test sand occurs after lowering the water head.

Figure VI.b.
Large column for infiltration experiments.

Figure VII.a.
Initial stage of PER spill into a trough containing simulated, moving groundwater.
Time = ~5 min.

Figure VII.b.
Spill of PER into a trough containing simulated, moving groundwater.
Time = ~20 min.

VII.a.

VII.b.

Figure VIII.a.
Spill of PER into a trough containing simulated, moving groundwater.
Time = ~1 h.

Figure VIII.b.
Spill of PER into a trough containing simulated, moving groundwater. Total spill volume ~24 L.
Time = ~2 h.

VIII.a.

VIII.b.

IX.a.

IX.b.

Figure IX.a. Closeup of bottom area of a PER spill into a trough containing simulated, moving groundwater. Spill volume ~24 L. Time = ~2 h. (Closeup of Figure VIII.b.)

Figure IX.b. Trough set-up used to study solubilization of a pool of CHC on the bottom of an aquifer.

X.a.

X.b.

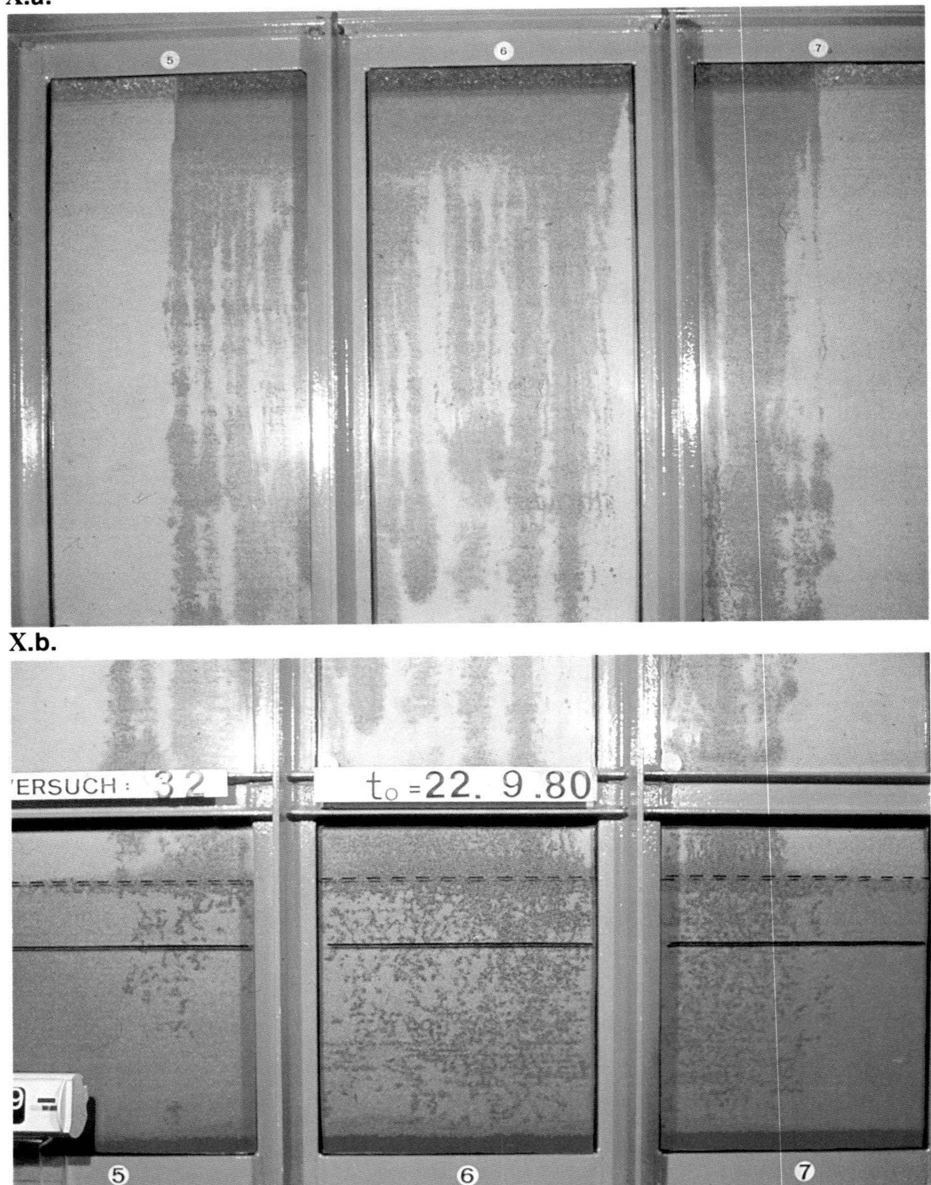

Figure X.a. Sheet-like spill of 36.3 L of PER. View of spill above the capillary fringe.

Figure X.b. Sheet-like spill of 36.3 L of PER. View of spill below the capillary fringe. Time = ~10 min.

XIII.a.

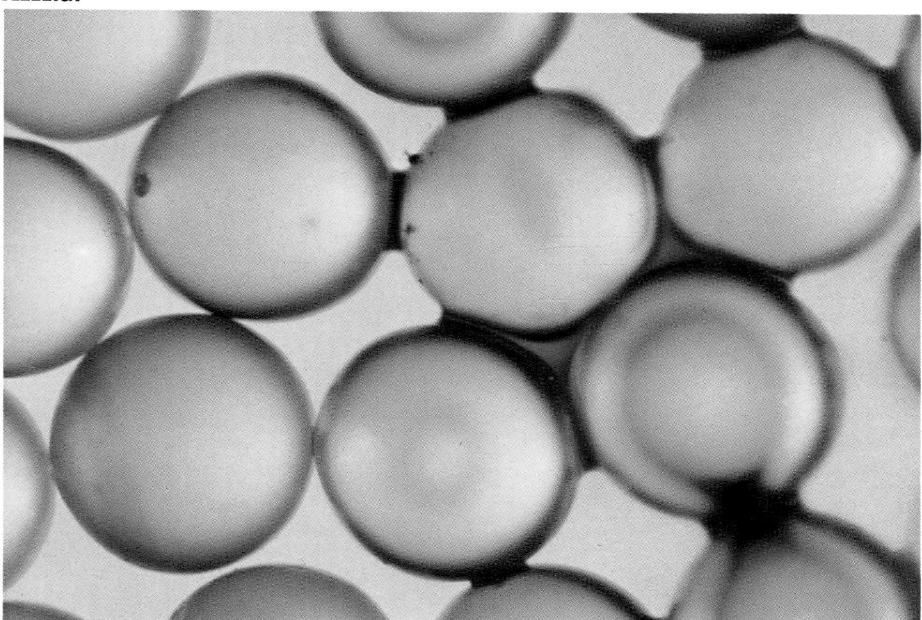

XIII.b.

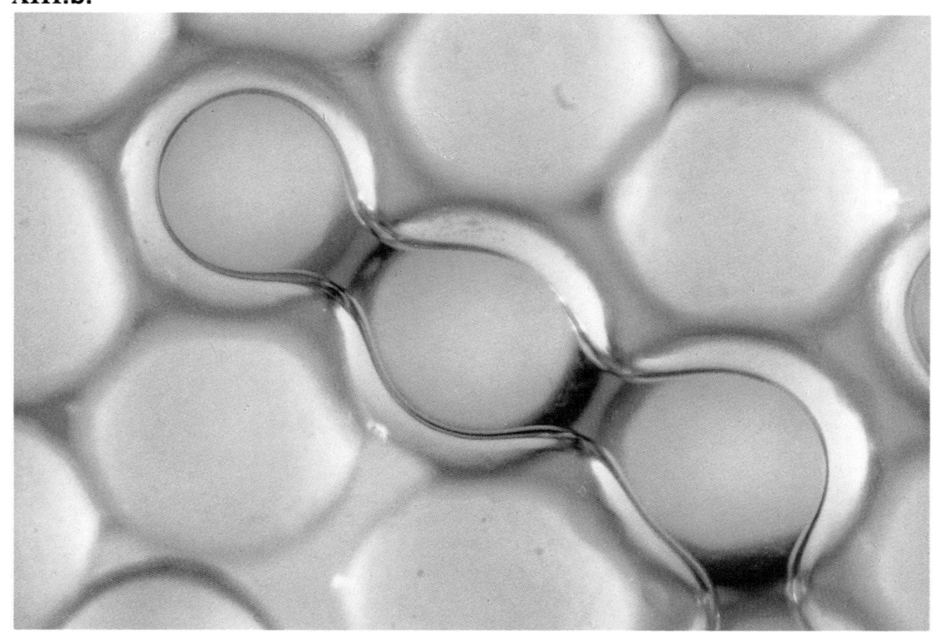

Figure XIII.a. Beads initially dry; water then dripped in from above. Smaller pores at right contain water.

Figure XIII.b. Beads initially dry; more water than in Figure XIII.a. was then dripped in from above. Numerous pores filled with water. Some bead/plate contact points wetted with water.

XIV.a.

XIV.b.

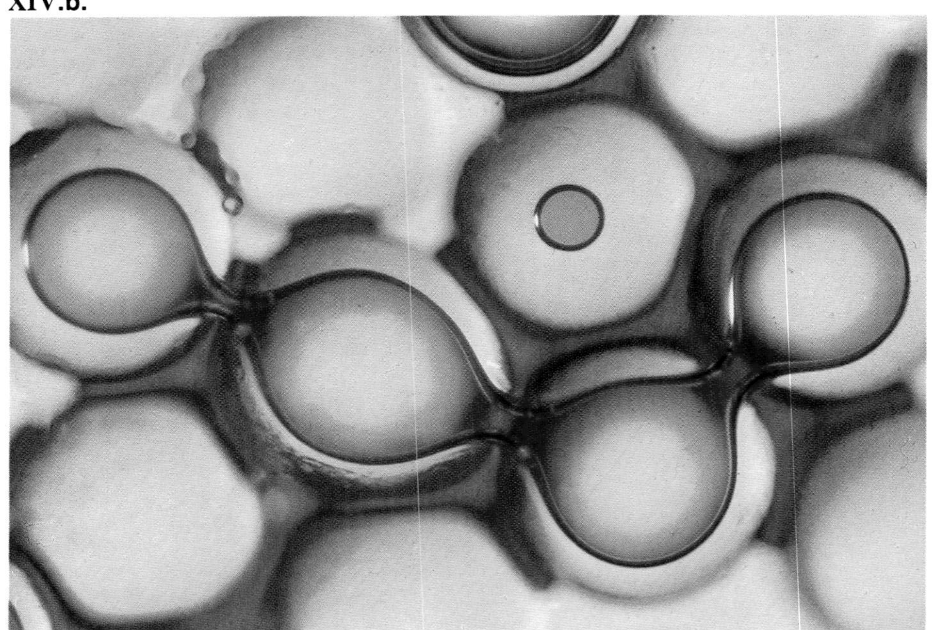

Figure XIV.a. Beads initially dry; PER then dripped in from above. Most of the pore corners contain PER.

Figure XIV.b. Beads initially dry; PER then dripped in from above. Some of the bead/plate contact points wetted and joined with PER.

XV.a.

XV.b.

Figure XV.a. Beads initially moist; diameter range = 0.85 - 1.23 mm. PER then dripped in from above. Most of the PER flowed out, leaving isolated drops in the internal portions of the pore spaces.

Figure XV.b. Beads initially dry; diameter range = 0.85 - 1.23 mm. PER then dripped in from above; it occupied the corners of the pore spaces. Beads then saturated from below with water; PER thereby forced into the smaller pore spaces.

XVI.a.

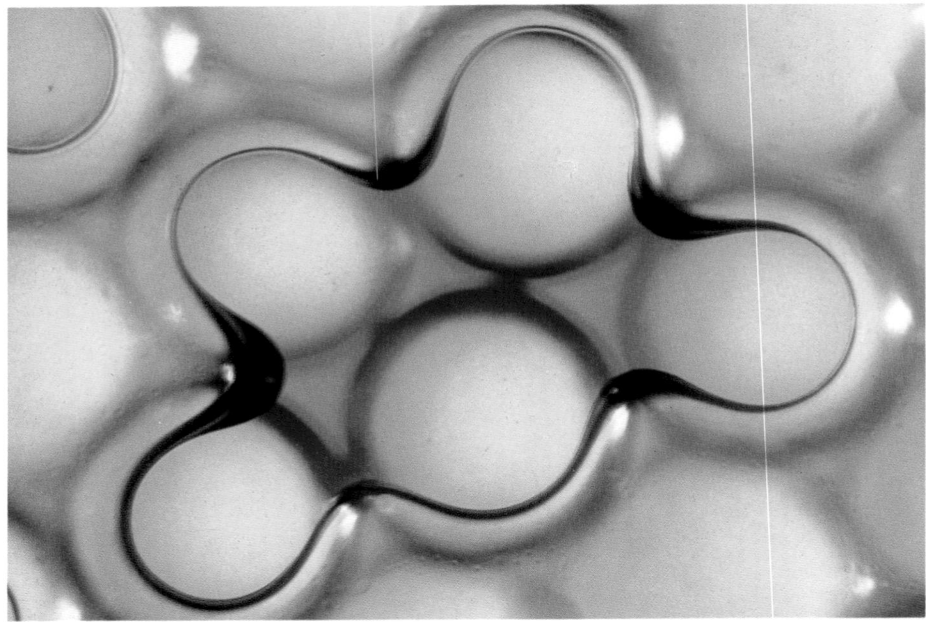

XVI.b.

Figure XVI.a. Beads initially moist; diameter range = 0.85 - 1.23 mm. PER then dripped in from above. The PER accumulated as a sheath around a zone of high water content.

Figure XVI.b. Beads initially saturated with water; diameter range = 0.49 - 0.70 mm. PER then applied from above. When the flow of PER was discontinued, the front portion of the PER stream broke off and halted.

XVII.a.

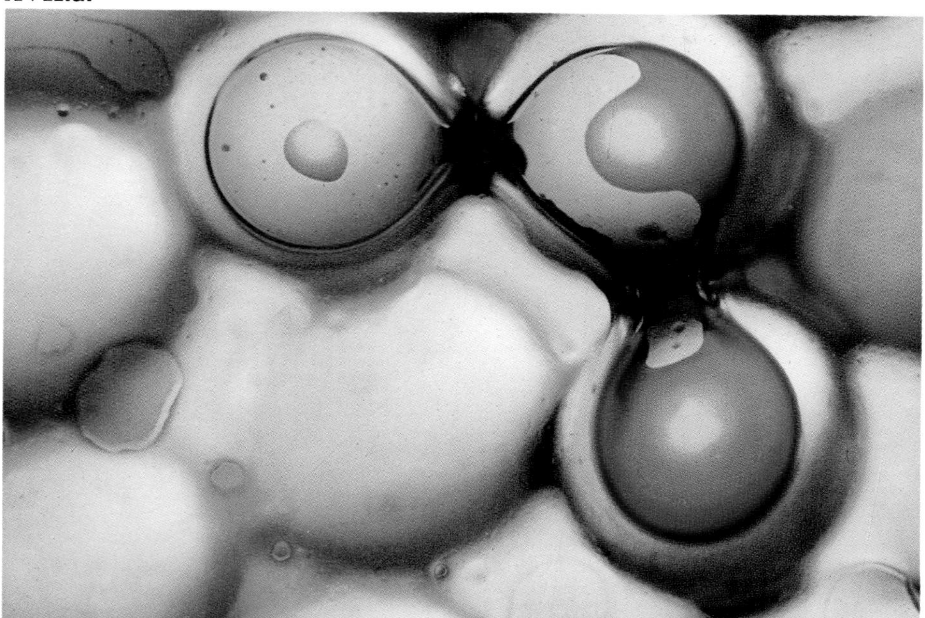

XVII.b.

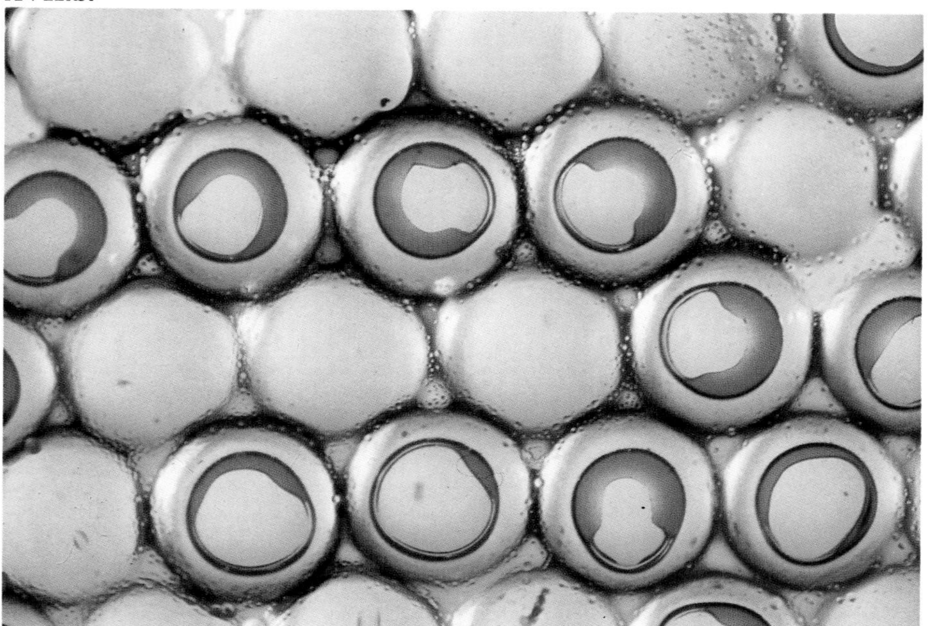

Figure XVII.a. Beads initially dry; diameter range = 0.85 - 1.23 mm. PER first applied from above. Water then applied from above. Infiltrating water drove PER from the bead surfaces.

Figure XVII.b. Later stage of process depicted in Figure XVII.a.

XVIII.a.

XVIII.b.

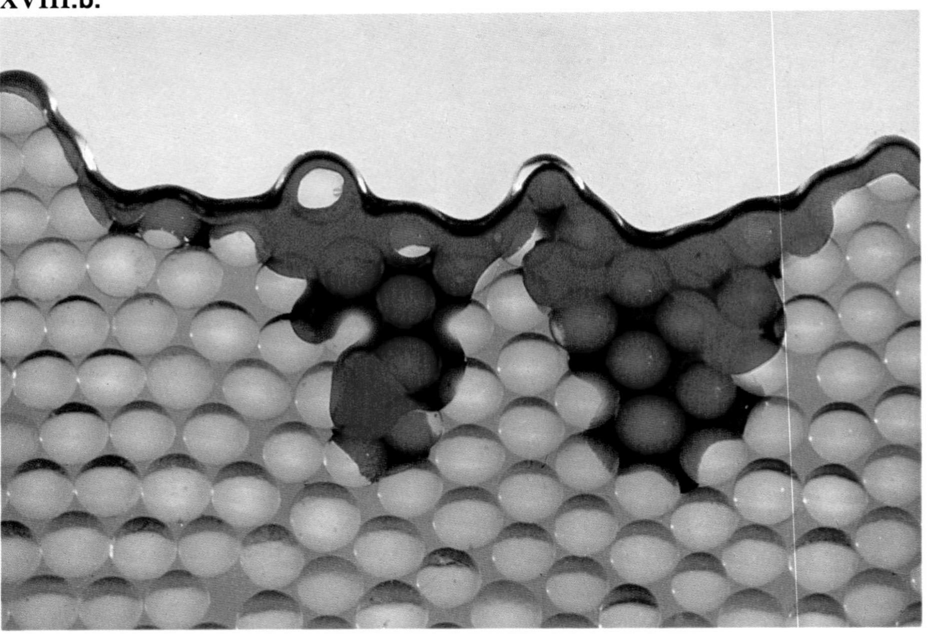

Figure XVIII.a. Beads initially saturated with water; diameter range = 0.85 - 1.23 mm. PER was applied and remained stable at the upper boundary of the capillary fringe.

Figure XVIII.b. More PER added to system shown in Figure XVIII.a. PER now is shown penetrating into the beads.

XIX.a.

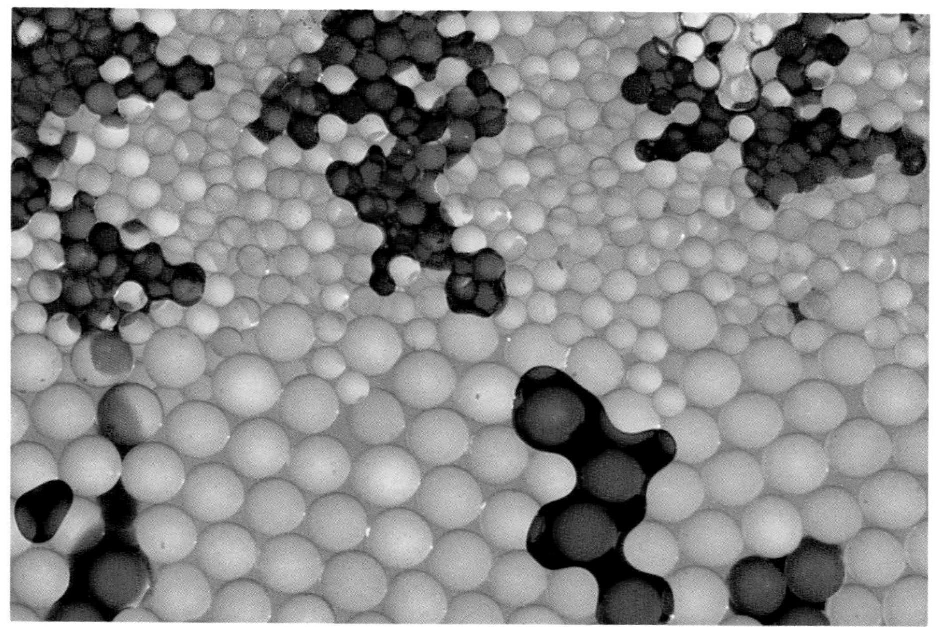

XIX.b.

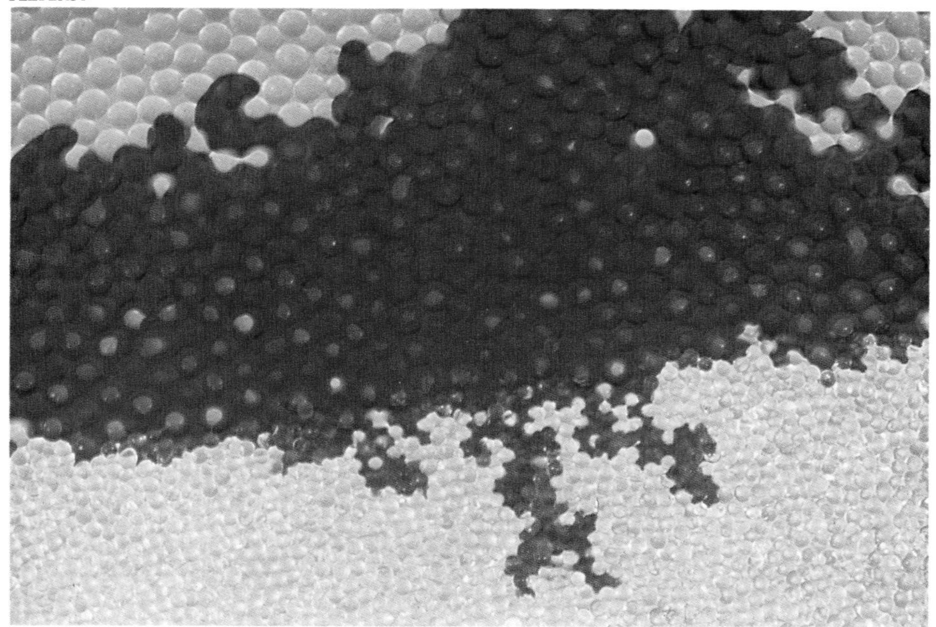

Figure XIX.a. Two-layer bead medium (0.49 - 0.70 mm diameter, upper; and 0.85 - 1.23 mm, lower) initially saturated with water. PER then applied from above. Infiltration paths are broader in the finer bead layer.

Figure XIX.b. Two-layer bead medium (0.85 - 1.23 mm diameter, upper; and 0.49 - 0.70 mm, lower) initially saturated with water. PER then applied. PER dammed up at interface with smaller beads, and entered them only when significant PER pressure built up in the upper layer. Disconnection of fingers occurred after application of PER ceased.

XX.a.

XX.b.

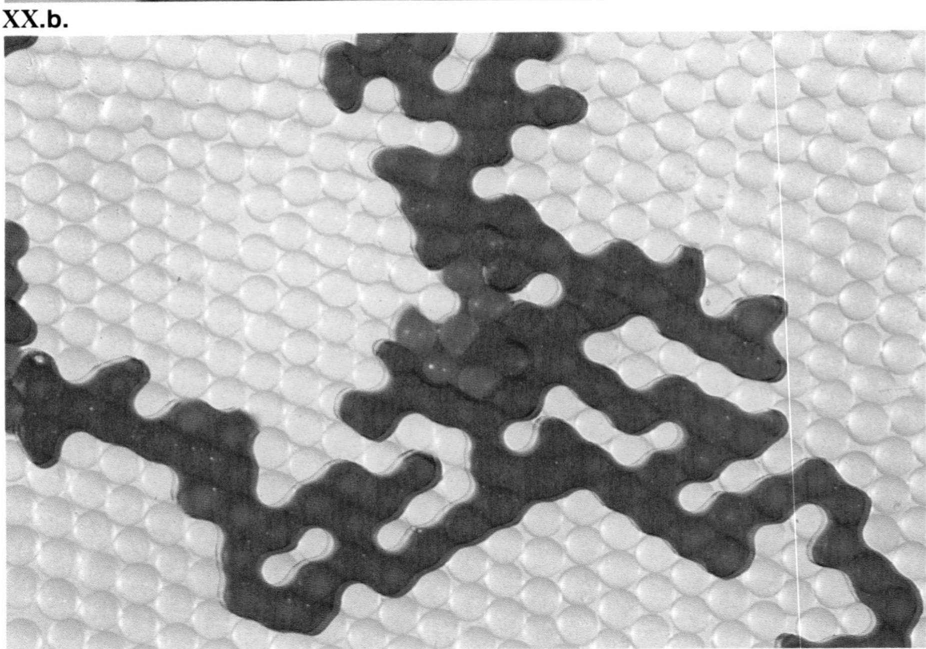

Figure XX.a. Beads initially saturated with water; diameter range = 0.85 - 1.23 mm. PER then applied from above. Finger of PER is shown detouring for no apparent reason around a "peninsula" of water.

Figure XX.b. Beads initially saturated with water; diameter range = 0.49 - 0.70 mm. PER infiltrated in a random manner.

XXI.a.

XXI.b.

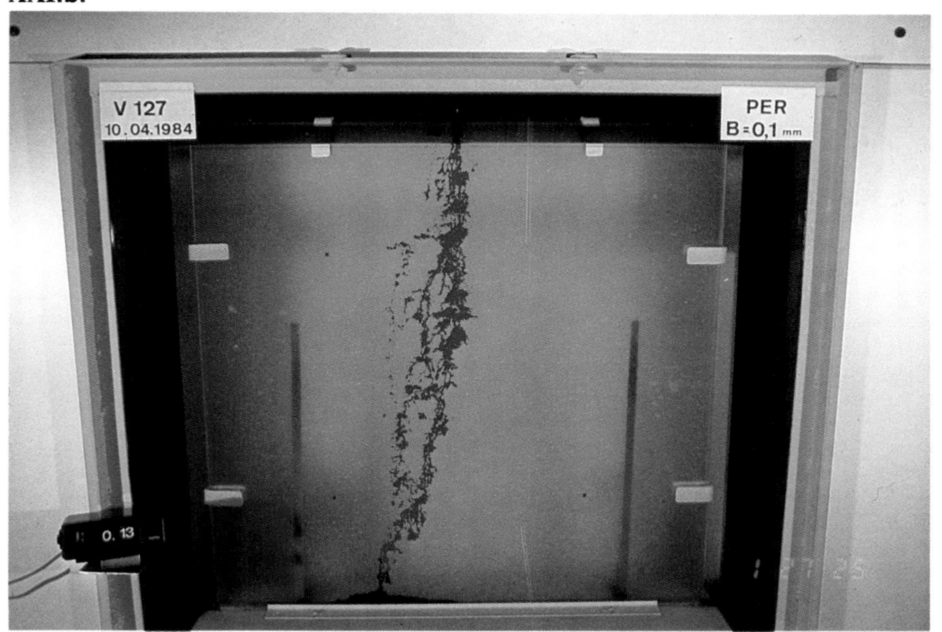

Figure XXI.a. Infiltration of PER into a fully saturated fracture with an aperture of 0.2 mm, and rough walls. PER applied at 1.4 mL/min.

Figure XXI.b. Infiltration of PER into a fully saturated fracture with an aperture of 0.1 mm, and rough walls. PER applied at 1.2 mL/min.

XXII.a.

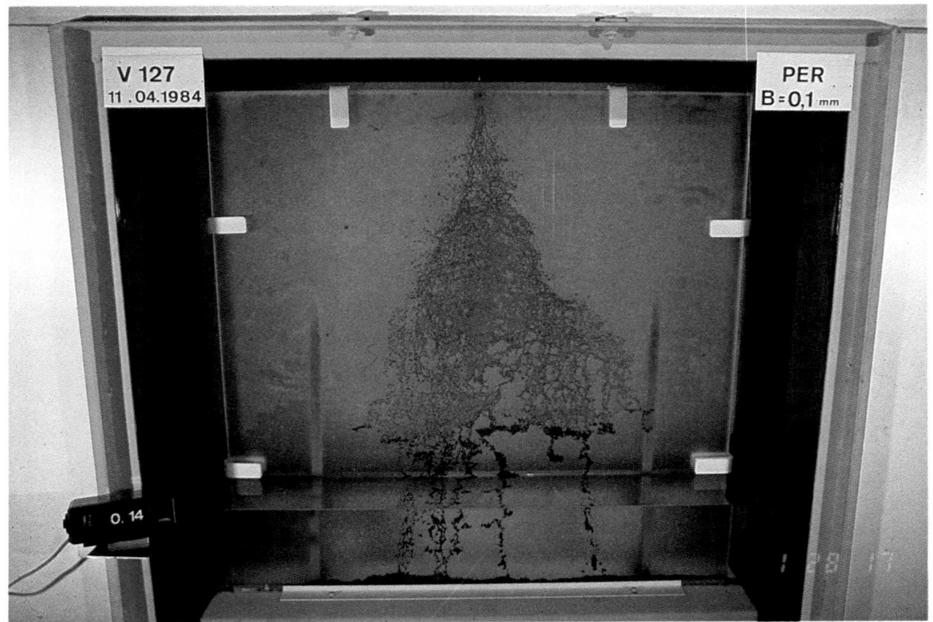

XXII.b.

Figure XXII.a. Fracture with an aperture of 0.1 mm, and rough walls initially saturated with water. Water level then lowered. PER then applied at 1.3 mL/min. Water remaining on fracture walls affected distribution of PER.

Figure XXII.b. Same system as in Figure XXII.a., but at 1.5 h after discontinuation of the application of PER. All PER shown is in a stable state of residual saturation.

XXIII.a.

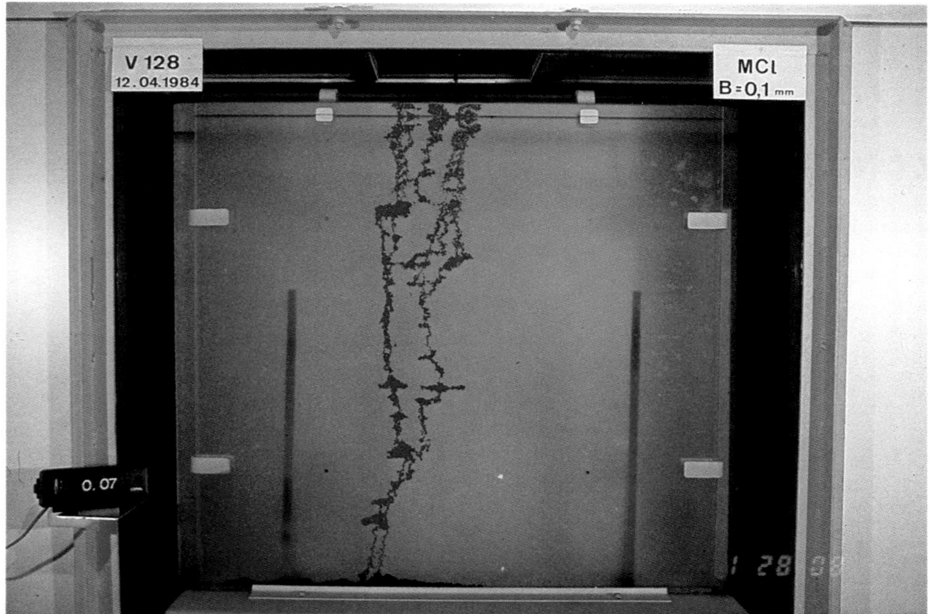

XXIII.b.

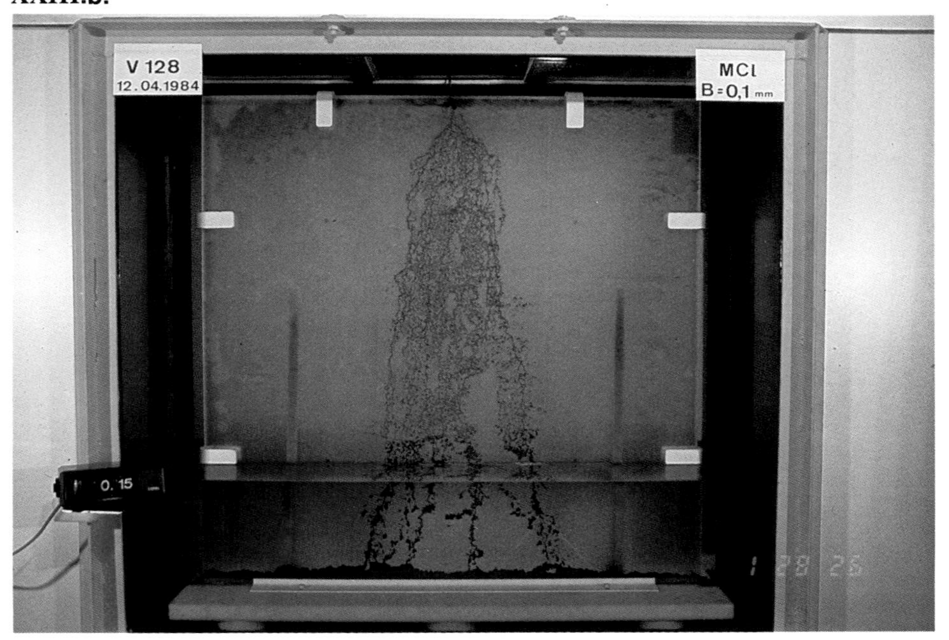

Figure XXIII.a. Infiltration of DCM into a fully saturated fracture with an aperture of 0.1 mm, and rough walls. DCM appled at 1.4 mL/min. Results similar to those for PER in Figure XXI.b.

Figure XXIII.b. Continuation of Figure XXIII.a. experiment. 45 min after ending the DCM application, the water level was lowered. Little DCM remained in the newly created unsaturated zone. 25 mL of additional DCM then added. Final results similar to those for PER in Figure XXII.a.

XXIV.a.

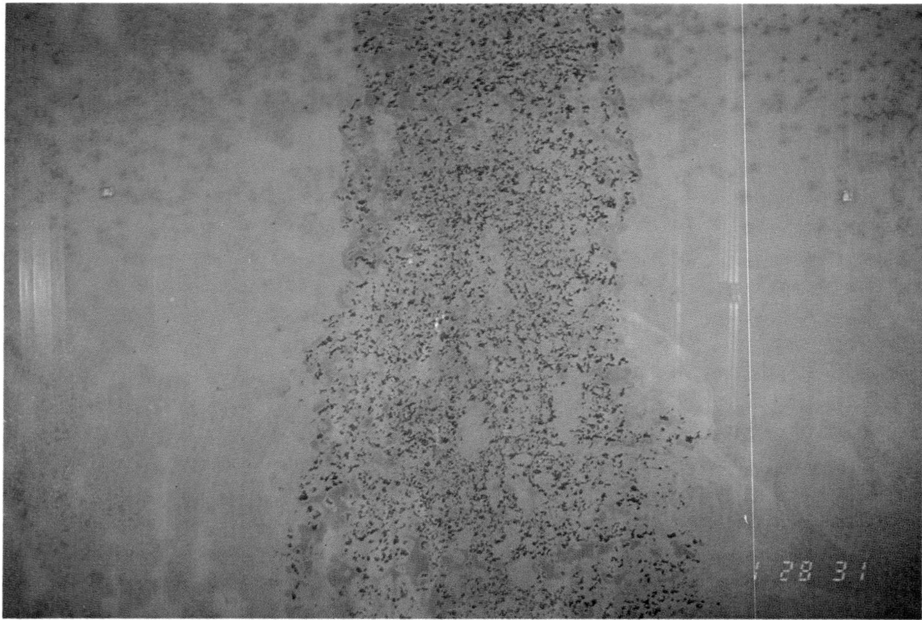

XXIV.b.

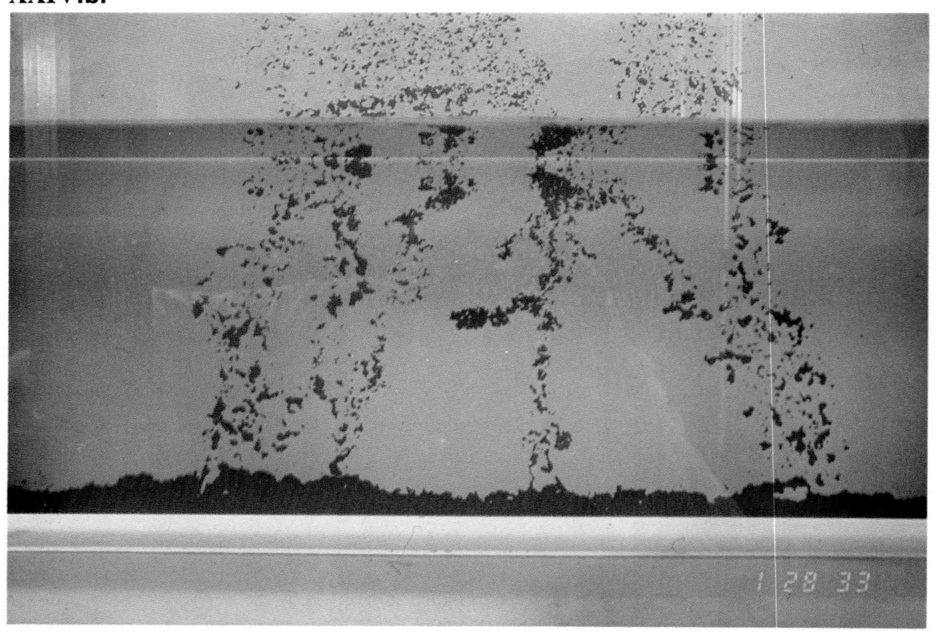

Figure XXIV.a.
Close-up of the unsaturated zone in the Figure XXIII.b. system 10 min after adding the 25 mL of DCM.

Figure XXIV.b.
Close-up of the saturated zone in the Figure XXIII.b. system 10 min after adding the 25 mL of DCM. The DCM on the bottom of the system is nearly free of water.

7

Migration in Aqueous Solution

7.1 DYE EXPERIMENTS ON DENSITY-AFFECTED FLOW IN POROUS MEDIA

Chlorinated hydrocarbons (CHCs) are fluids which are immiscible with water. Nevertheless, they do dissolve in water to a small extent. We recognize that a given volume of CHC cannot spread beyond a certain volume (or length, in a one-dimensional sense) of porous medium. It is therefore important to determine the extent to which a CHC will be removed from a spill volume by dissolution into both infiltrating precipitation and moving groundwater. The rate and manner in which a CHC is so removed will determine the final overall dimensions of the zone of contamination.

The spreading of dissolved materials and the nature of their concentration distributions will be determined primarily by the flow processes that occur in the pore spaces, i.e., by events on the microscopic scale. The conceptualization of these flow processes forms the basis of the hydrodynamic dispersion model, as well as the methods by which hydrodynamic dispersion is measured. We have maintained for many years that the effect of hydrodynamic dispersion is largely overestimated. A much more important role is played by the convection processes that occur due to the heterogeneities present in most aquifers. Therefore, the dispersion parameters obtained with the aid of field observations are, in reality, fictitious values. They are obtained when one views the aquifer *as if it were a homogeneous medium*. If a concentration distribution is predicted assuming homogeneity, then discrepancies between the prediction and reality will generally be observed.

Although we have worked in the area of hydrodynamic dispersion in porous media for over a decade, this topic will not be discussed here; the Institute for Hydraulics at the University of Stuttgart has already reported in detail on this subject within the context of the Baden-Württemberg CHC Research Program. We must nevertheless provide a discussion of the effect of the density of a solution on the vertical distribution of solute transported in an aquifer. This issue is of particular importance in the spreading of CHC. However, the perceptions of researchers on this particular issue are divided, and many schematic diagrams have been drawn without the benefit of any experimental results; the potential exists for the persistence of many incorrect conclusions.

Several experiments pertaining to this issue were carried out using the model trough diagrammed in Figure 3. The hydraulic advantages of an inclined trough have already been discussed in Section 3.2. As discussed in Section 3.2, in order to complement the perchloroethylene (tetrachloroethylene, or PER) experiments described there (Experiments I and II), water dyed with fluorescein at 1–2 g/L was infiltrated into the trough. A saturated solution of 1,1,1-trichloroethane (1,1,1-TCA) will have a concentration similar to this and, therefore, will also have a similar density, i.e., 1 to 2 g/L greater than the water flowing through the trough. The dyed water was added at the same place the PER was added in Experiment I (Figure 5). The infiltration rate was 15.3 L/hr, i.e., 3.3 L/hr greater than in the PER experiment. All other experimental conditions were the same. The character of the spreading of the dyed water observed was presented in Figure 4.

Initially, the dyed solution sank in the unsaturated zone and penetrated (by virtue of the high head) into the lower layer of the simulated two-layer aquifer. The groundwater flow quickly forced the solution to move horizontally. The relative rates of movement in the two layers were in accordance with the differing hydraulic conductivities of the layers.

Similarly, the subsequent complete removal of the dye by the groundwater flow occurred more quickly in the lower layer. While the dyed solution possessed a density that was greater by 1 to 2 g/L over that of the model groundwater, it did not demonstrate any measurable tendency to sink.

In order to examine the effects of lateral dispersion and density more closely, stationary injection tubes were mounted transverse to the sidewalls of the same model trough (Figure 34). A 0.35-g/L fluorescein solution was added at a very low rate in order to simulate a line source as closely as possible. The hydraulic gradient was 0.0036. Neglecting the slight downward movement that occurred in the first meter, the dye solution did not sink in any meaningful amount (Figure 35). The initial downward trend was presumably due to causes other than density. The dispersion in the vertical direction was negligible; the very narrow plumes remained separated even over a 5-m travel distance.

Further experiments were carried out with the same model trough but using a different sand. The hydraulic gradient was 0.0025, and the fluorescein concentration was 2 g/L. The results were presented in part in Figures Ib, IIa, and IIb. The behavior of the two dye traces further substantiated the earlier observations. In contrast, subsequent experiments with dye concentrations of 5 and 10 g/L demonstrated a noticeable initial sinking of the dye solution. This sinking decreased markedly within a short distance of the injection line. Although the spreading that resulted was considerable, it

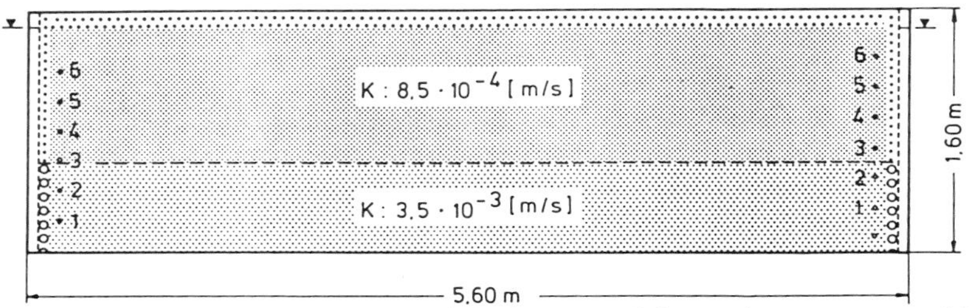

Figure 34. Schematic diagram of the large model trough as it was used in the dispersion experiments.

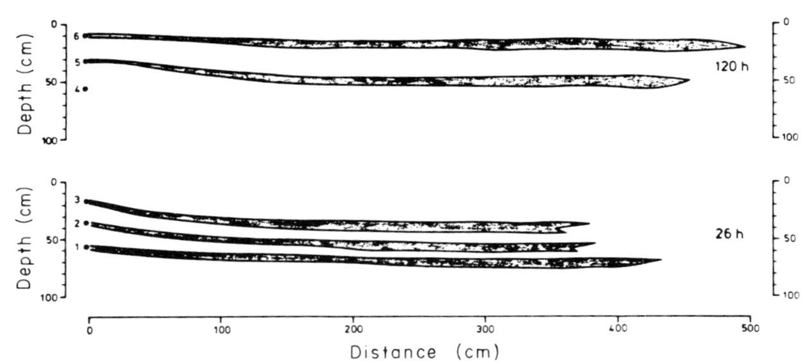

Figure 35. Results of the dispersion experiment with a 0.35-g/L fluorescein solution.

only occurred in the downward direction. In 5 m of travel distance, the vertical spreading observed was 0.5 m at a dye concentration of 5 g/L, and 0.8 m at 10 g/L. It should be noted that with both dye solutions, the upper boundaries of the traces were parallel to the water table; they became slightly misshapen only with increasing distance from the injection line. Over the same interval, the bulging lower boundary sank substantially. The dispersion process could thus be described as one-sided in the negative z direction with no dispersion occurring in the positive z direction.

The water solubilities of most CHCs are less than 2 g/L. This means that even with saturated solutions of these CHCs, the density of contaminated water will not be sufficient to cause substantial sinking of a plume. Since the solubilities of dichloromethane (DCM) and chloroform are of the order of 10 g/L, however, their saturated solutions will be subject to substantial sinking. In most cases, the concentrations near all CHC spill sites are very low – usually far below the saturation values. This indicates that it may be assumed that density-affected flow will be the exception in real-world situations. The primarily horizontal character of the dissolved plumes in Figures IIIb, IVa, and IVb has therefore been represented correctly.

7.2 DYE EXPERIMENTS IN FRACTURED MEDIA

We have investigated the spreading of low-concentration solutions ($\leq$ 2 g/L) in fractured rock for several years. This has been done with the help of a model trough that allowed the simulation of two perpendicular vertical sets of fractures. (See F. Schwille and K. Ubell, "Flow Processes in an Interconnected Fracture Model in the Vicinity of Pumping Wells," [in German], *Deutsche Gewässerkundliche Mitteilungen* *29*(1):13–26, 1985.) The sum of the lengths of the fractures was 52 m. Numerous experiments with this system have shown that density effects are more important in fractures than in porous systems. Nevertheless, with solutions of <1 g/L, no significant tendencies toward sinking were observed.

These experiments also showed that the process of spreading in fractured media is much different than that in porous media. In most cases, one cannot satisfactorily predict spreading in fractured systems using the known formulas for hydrodynamic dispersion as expressed in terms of longitudinal and transverse dispersion coefficients. In other words, fractured media usually cannot be modeled as quasi-equivalent porous media. As a result, in this work, the manner in which the fracture aperture controls the flow process in a fracture net was investigated with a mathematical model. The results obtained from the mathematical model were in agreement with those from the physical model.

7.3 SOLUBILIZATION: REMOVAL OF CHCs AT RESIDUAL SATURATION IN POROUS MEDIA

At the conclusion of the spreading process, CHC bodies will be immobile and will be present in a state of residual saturation. Although the water flow velocity can be reduced substantially, these CHC bodies will be permeable to precipitation infiltration and groundwater flow. The basis of their permeability is evident in the photos obtained in the experiments carried out on the microscopic scale. Those photos also illustrate that the distribution of CHC into droplets, rings, grain-encircling "skins," and films serves to create a very large contact surface area. The components of the CHC body will dissolve from that surface area, then diffuse into the surrounding water. When the water flow rate is very low, the water can become very nearly saturated after passing only a short distance through a CHC body. Downgradient of the body, "strands" of dissolved CHC will form in a manner that depends upon the nature of the spill. These strands, however, will usually only be detected as some type of average when water samples are taken for analysis; often only an average concentration will be obtained even when a distinct vertical profile exists.

The ability of groundwater to remove CHC by dissolution was first investigated with a very simple experimental setup (Figure 36). Glass columns were filled with different experimental sands. Three different CHC compounds were investigated. For each column, 100 mL of a CHC dyed with oil red was added to create a state of residual saturation. Tap water was percolated through each column. The effluent concentration and the cumulative mass of solubilized CHC were followed as a function of the water volume (Figure 37). The percolation was carried out with the flow from bottom to top so as to avoid promoting a sinking of the CHC phase. The flow velocity was approximately 1 m/day.

In no case did the water physically displace a CHC; the effluent water was found to contain only dissolved CHC. In spite of the fact that the layer of sand in CHC residual saturation was thin, PER was initially found to be present at saturation in the effluent. For trichloroethylene (TCE) and 1,1,1-TCA, initial concentration values between 70 and 80% of saturation were obtained. The less than 100% saturation values for TCE and 1,1,1-TCA were presumably the result of unavoidable volatilization losses during sampling. The CHCs were solubilized from the columns until the remaining

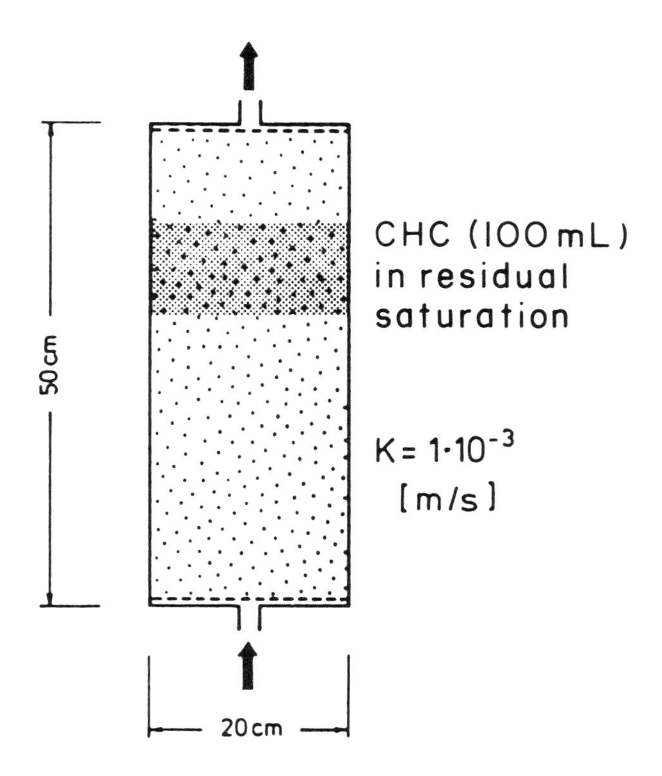

Figure 36. Schematic diagram of the simple solubilization experiment.

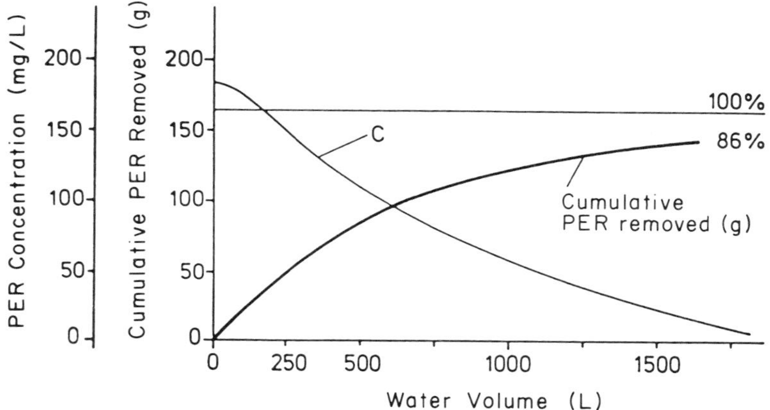

Figure 37. Results of the solubilization experiment.

CHC levels were sufficiently low that further removal was slow.

The apparatus in Figure 38 was used for the subsequent solubilization experiments. The glass column was filled with 90 cm of sand and then infiltrated with CHC to near the residual saturation. The remaining 10 cm of column was then filled with sand and a filter layer in order to collect droplets of CHC that might become mobilized by the upward-moving water. The column was then sealed with its cover plate, flooded, then percolated at a constant rate with degassed, 10°C water. The flow rate was set at about 1 m/day. Since the pore volume of the column was 12 L, the volume flow rate was 0.5 L/hr. The effluent water was collected in a separatory funnel. The inflow tube to the funnel was placed at the bottom of the funnel. In this manner, water that initially flowed into the funnel formed a protective layer for the water that later passed into the funnel. Volatilization losses were thereby largely avoided.

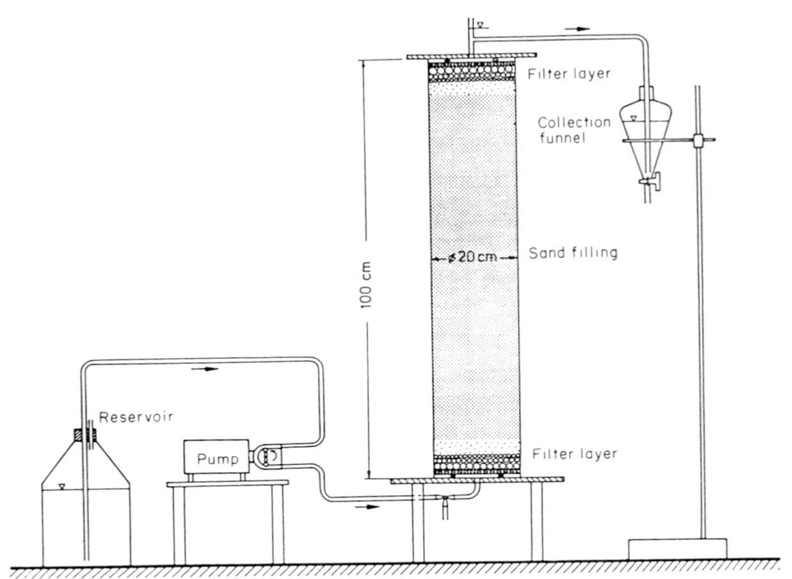

Figure 38. Standard apparatus for the study of solubilization.

The solubilization of PER was studied as a function of the grain size (i.e., the pore size) and the hydraulic conductivity K. Three different experimental sands were used. The average concentrations obtained are shown in Table 3.

The concentration of the solubilized PER was not found to depend strongly on the grain size (i.e., on the K). The concentrations found were higher than the commonly cited solubility of 160 mg/L; perhaps the solubility at the experimental temperature of 10°C is higher than the handbook value at 20°C. [Translators Note: C. T. Chiou, V. H. Freed, D. W. Schmedding, and R. L. Kohnert (*Environ. Sci. Technol.* 11:475–478, 1977) have reported a PER solubility of 200 mg/L at 20°C. See Appendix I for more information.]

In any event, these experiments have demonstrated that water will become saturated with PER after passing at 1 m/d through only 1 m of a porous medium that is at residual saturation with PER. Therefore, it may be assumed with some certainty that only a relatively short contact time is required to produce high CHC concentrations in infiltrating precipitation and groundwater, even when those waters are moving at moderate velocities. The commonly used method of artificial infiltration for removing CHCs from the ground is thereby found to be fundamentally sound. The extent to which intensive pumping will accelerate the decontamination has not yet been investigated. However, as the following experiments demonstrate, increasing the water flow rate also causes an increase in the total CHC mass transport rate. It is our recommendation that each remediation operation be carefully optimized.

Table 3. Average Concentrations of Solubilized PER with Three Experimental Sands

Sand K (m/sec)	Concentration (mg/L)
2×10^{-4}	198
2×10^{-3}	219
6×10^{-3}	204

7.4 SOLUBILIZATION: REMOVAL OF CHCs IN POOLS

After a CHC has sunk through an aquifer, it will spread itself on the bottom of the confining layer in a manner that will depend upon the slope of that layer, as well as upon other characteristics of the particular system. The CHC will collect in pools and puddles wherein the degree of saturation will be relatively high. As the groundwater passes through the porous medium above the CHC zones, the CHC will dissolve from the CHC surfaces and diffuse into the groundwater. In time, the groundwater will carry away the CHC. This process will proceed at a rate substantially slower than if the same amount of CHC was dispersed in a residual saturation situation, and the groundwater was moving past the high surface area present under those conditions.

In order to obtain an understanding of the order of magnitude of the removal rate from pools on the bottom of an aquifer, we constructed a flat glass trough 150 cm long and 50 cm wide (Figure IXb). The goal of the experiments was to create a defined pool of CHC on the bottom of the trough over which the water would flow. The use of glass to construct the trough was considered essential since it was necessary to determine whether or not a pool of CHC of the desired form had been created. Two low sills made from angle iron were installed on the bottom of the trough. They were separated by a distance of 100 cm. The sills are visible as black lines in Figure IXb. The trough was filled with a uniform sand with a good hydraulic conductivity ($K = 3.5 \times 10^{-3}$ m/sec). The first experiments ended disappointingly inasmuch as it did not prove successful to bond the sills to the bottom of the trough with CHC-resistant glue; after a certain time, the CHC spread itself over the entire bottom surface. The front edge of the pool is visible in Figure IXb. We later abandoned the use of sills and outfitted the trough as outlined in Figure 39.

The addition of the CHC was made at the bottom and

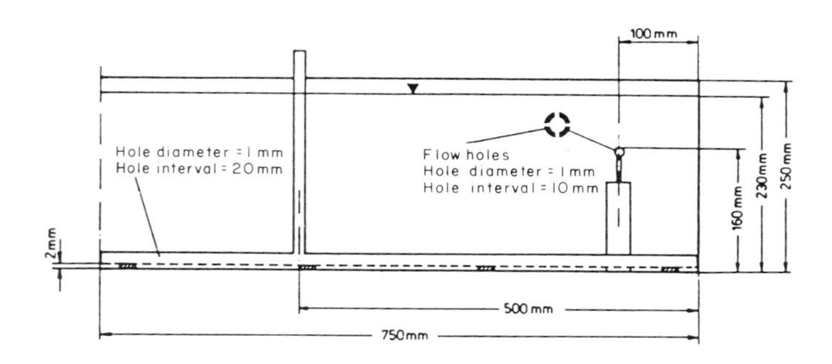

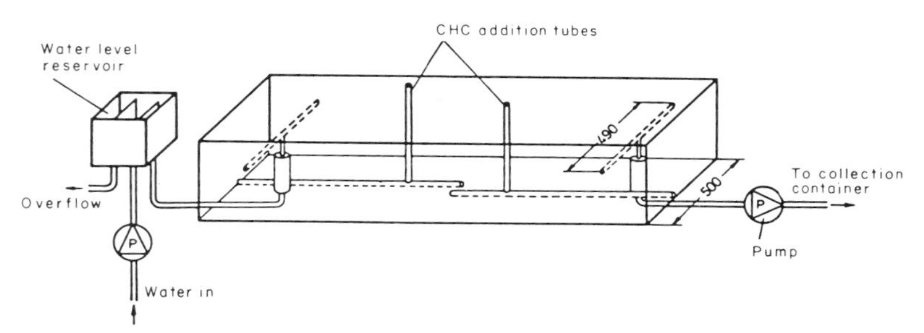

Figure 39. Schematic diagram of the flat trough for the determination of the removal of CHC from pools.

along the long axis of the trough through small perforated tubes. The addition and removal of the moving water for the trough was made through perforated tubes that were installed perpendicularly to the long axis of the trough at a height of 16 cm above the bottom. The addition of the CHC proceeded without difficulty. We were convinced, by virtue of the previous experiments, that each CHC layer created covered the entire bottom of the trough and did so with a uniform thickness.

1,1,1-TCA was selected as the CHC for the first experiment since both its vapor pressure and its solubility represent intermediate values for the solvents. TCE was selected for the second experiment. The water table was positioned 27 cm above the bottom of the trough. Pumps were used to produce water flow through the trough since they allow a closer regulation of the flow than is possible by simply setting the hydraulic gradient with reservoirs. The trough was covered tightly with a glass sheet in order to minimize volatilization losses. The room temperature was 20 to 22°C.

The results from both experiments are summarized in Table 4. The data has been arranged in order of increasing water flow rate to facilitate the consideration of the results. The total rate of CHC removal from the trough also follows this order. The actual experimental order is given by the run numbers (column 1). Taken by themselves, the volume flow rates (column 2) have little meaning. However, together with the total pore volume, they allow the calculation of the flow velocities (column 4). Considering that the flow in the trough was not completely simple, the flow velocities must be used with some care. Each CHC concentration (column 3) is the average of the values obtained with its corresponding volume flow rate. With 1,1,1-TCA, substantial fluctuations in concentration were observed. It is possible that these fluctu-

Table 4. Results of the Solubilization Experiments in the Flat Trough

1	2	3	4	5	6	7
Run Number	Volume Flow Rate (L/hr)	CHC Concentration (mg/L)	Water Flow Velocity (m/day)	Removal Rate for the Total Trough Area (g/day)	Theoretical Removal Time (days)	Removal Rate for 1 m^2 (g/day)
I - 1,1,1-TCA[a]						
3	1.5	170	0.7	6.1	447	8.1
1	4.5	89	2.0	9.6	284	12.8
2	14.8	85	6.7	30.1	91	40.1
II - TCE[b]						
1	1	90	0.45	2.2	1690	2.9
6	2	67	0.9	3.2	1137	4.3
2	2	87	0.9	4.2	873	5.6
3,4	4	73	1.8	7.0	521	9.3
5	6	77	2.7	11.1	329	14.8

[a]Total amount 1,1,1-TCA added = 2728 g; total amount removed = 214 g = 7.8%.
[b]Total amount of TCE added = 3650 g; total amount removed = 190 g = 5.2%.

ations were inherently caused by the experimental design. Nevertheless, allowances could be made for them, thereby permitting the appropriate conclusions to be drawn. In contrast to 1,1,1-TCA, TCE did not display concentration fluctuations within each volume flow rate.

As with the volume flow rate data, the concentration data are of little value in and of themselves. However, together with the volume flow rate data, one can calculate the removal rate for the whole trough (column 5). The theoretical time required to remove all of the material from the trough (column 6) may be obtained from the column 5 data and the total mass added to the trough. The daily removal rate for a 1-m^2 surface area is presented in column 7; strictly speaking, these rates pertain only to this 1.5-m × 0.5-m trough. Overall, the data in columns 4 and 7 suffice for a consideration of the results in Table 4.

Both series of experiments (I and II) gave similar results. As the flow rate increased, so also did the removal rate. While the two parameters were not directly proportional to one another, there was a nearly linear relationship. Also, as expected, the removal rates for 1,1,1-TCA were higher than for TCE. Other than the above, however, the results were not of adequate detail to allow the determination of the functional relationships governing CHC removal under these conditions.

Since the total rate of movement of material across the CHC/water interface increases with increasing water flow rate, it will be possible to accelerate the decontamination of the bottom of an aquifer by increasing the pumping rate at a well. However, at some point, further increases in the associated financial and water resource depletion costs will not be supportable. It will always be necessary to try and reach the optimal cost-benefit ratio. Since the diffusion coefficients of CHC depend upon the compounds themselves, as well as upon the nature of the porous medium, the removal rates will depend upon the CHC and upon the type of soil. Therefore, for the same type of soil system, DCM will certainly be removed much faster than will PER. For the same CHC compound and the same hydraulic gradient, the removal will occur more quickly in a more permeable medium (e.g., the lower level gravels of a flood plain) than in a less permeable medium (e.g., the fine sands of an intermediate alluvial layer). Finally, the importance of the thickness of the CHC layer cannot be overlooked; a layer that is 2 to 3 cm thick (as used in our experiments) will be removed more quickly than a 10-cm layer.

The above considerations lead to the conclusion that CHCs can, in extreme cases, persist on the bottoms of undisturbed aquifers for time periods ranging between several months to decades. The theoretical removal times (Table 4, column 6) permit an understanding of the orders of magnitude of these lifetimes.

7.5 ADSORPTION-DESORPTION IN POROUS MEDIA

Since they are lipophilic, CHCs will be adsorbed in significant amounts to soils that are high in organic matter. For information on the adsorption-desorption of CHCs to soils containing organic matter, the reader is referred to the reports of the Engler-Bunte Institute; this document will deal with the behavior expected in nonsorbing soils.

For soils that contain nearly pure minerals and little organic matter, the literature indicates that CHCs will not sorb to any significant degree. Most high-yield aquifers are nearly free of organic matter. This is particularly true for the coarse-grained aquifers. Thus, when dissolved CHCs reach coarse layers, they will be nearly unretained. This is important since the transport of solutes is largely determined by what happens convectively in the coarse layers. Therefore, since CHCs other than DCM exhibit little biodegradability in the subsurface, the potential for concentration reduction with increasing distance from the source is often limited to what can occur by hydrodynamic dispersion.

In order to test the working theory that CHCs do not sorb on mineral soils, we undertook percolation experiments with quartz sands that were low in clay. Such experiments are more representative of what happens under natural flow conditions than are batch experiments. The flow rate was selected so that it was representative of the flow velocities which occur in natural groundwater aquifers. Through the use of selected model sands and by the application of a tested experimental technique, it was possible to produce a very homogeneous medium, a uniform flow, and an extremely sharp breakthrough front. For all intents and purposes, the experimental conditions approached ideality. The apparatus (Figure 38) was the same as we have used to obtain longitudinal dispersion coefficients.

In the following examples (V1 to V4), the porous medium was comprised of a very uniform, intermediate sand (Figure 1, Curve 2). The most important experimental data are listed in Table 5.

After being flooded from below, each cylinder was rinsed for several days with deionized and partially degassed water. A flow velocity of 1 m/day was used for the actual adsorption experiments. Therefore, each day, one pore volume was replaced. Samples of 0.5 L were taken on an hourly basis. The main concern was to avoid volatilization losses.

In order to provide an easily determined, conservative tracer, 10^{-3} *F* (FW/L) NaCl was added to each CHC solution. The NaCl concentration was determined by conductivity. The CHC analyses were carried out by extracting with pentane, then injecting into a gas chromatograph.

The experimental results are presented in Figure 40 in the

Table 5. Experimental Data

Total pore volume of the medium V_p	13.1 L
Porosity	42.2%
Hydraulic conductivity, K	$\sim 1 \times 10^{-3}$ m/sec
Water uptake	11.6 L
Degree of saturation of the pore space	88%

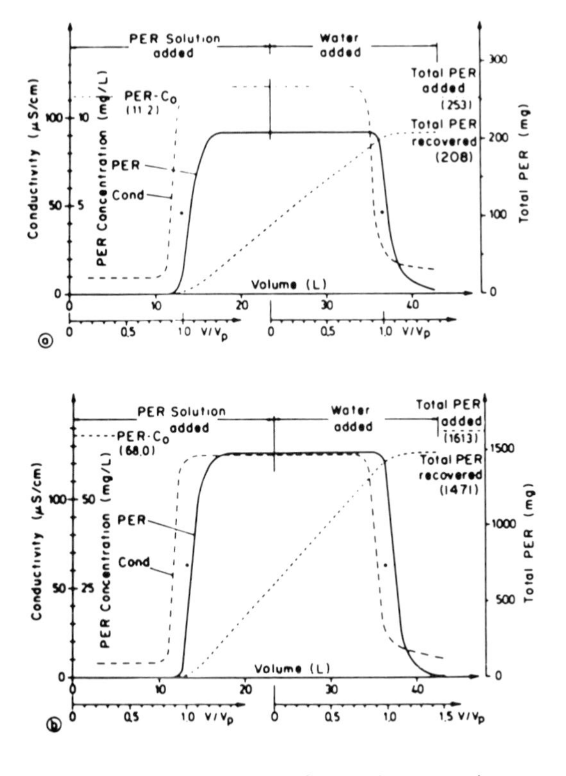

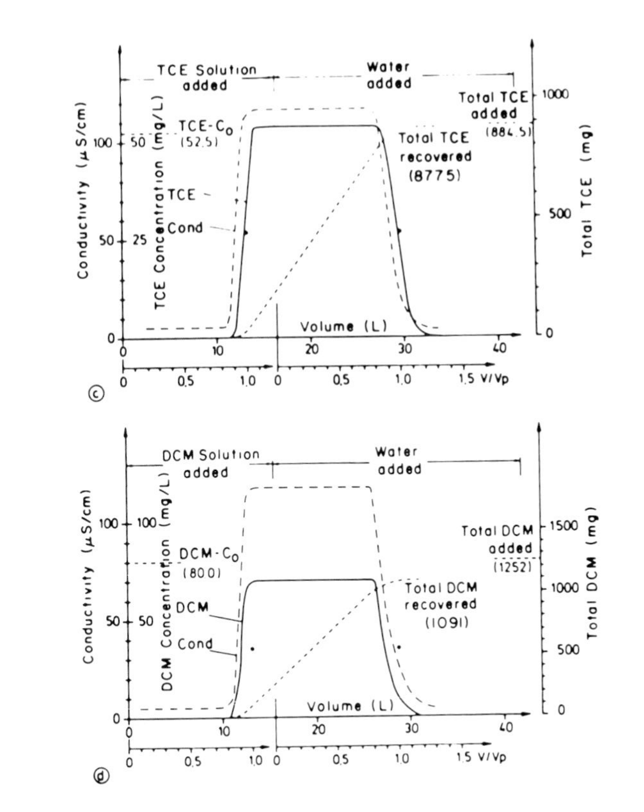

Figure 40. Adsorption-desorption experiments with PER, TCE, and DCM in a low-clay content quartz sand.

form of breakthrough and cumulative sum curves. In order to present the concentration vs volume results in a form which is normalized to the experimental conditions, the additional scale of V/V_p is given for the abscissa. This was accomplished by dividing the volume V by the calculated pore volume V_p. Dots have been used to mark the points with the coordinates ($C/C_0 = 0.5$, $V/V_p = 1.0$). In the plots, the total amounts of CHC recovered are compared to the total amounts of CHC added. Table 6 presents the volumes whereupon breakthrough (BT) first took place as well as the volumes required to achieve $C/C_0 = 1$. The volumes required to achieve $C/C_0 = 0$ and breakthrough of the CHC-free water are also included. The last two quantities were measured from the point at which CHC-free water was substituted for the initial, CHC-containing water. The state of increasing CHC concentration and decreasing concentration are denoted in Table 6 as (↑) and (↓), respectively.

The experiments allowed the following conclusions to be made. For NaCl, it may be assumed with confidence that the chloride was not sorbed. For all of the experiments, the

Table 6. Breakthrough (BT) Volumes, Volumes to Achieve $C/C_0 = 1$, and Volumes to Achieve $C/C_0 = 0$ under Increasing (↑) and Decreasing (↓) Concentration based on Conductivity and CHC Concentration.[a]

		Conductivity			CHC		
		BT	$C/C_0 = 1$	$C/C_0 = 0$	BT	$C/C_0 = 1$	$C/C_0 = 0$
V1	↑	0.75	1.03		0.87	1.32	
(PER)	↓	0.78		1.43	0.95		1.43
V2	↑	0.73	1.06		0.89	1.23	
(PER)	↓	0.81		1.47	0.97		1.47
V3	↑	0.81	1.04		0.84	1.11	
(TCE)	↓	0.79		1.24	0.84		1.32
V4	↑	0.75	1.02		0.79	1.02	
(DCM)	↓	0.79		1.29	0.79		1.25

[a]All Volumes as V/V_p

breakthrough of chloride began within the range $V/V_p = 0.73$ to 0.81. The maximum concentration was reached within $V/V_p = 1.02$ to 1.06. The breakthrough of the CHC-free water occurred within $V/V_p = 0.78$ to 0.81. After an additional 1.24 to 1.43 pore volumes, the conductivity decreased to the initial value of ~5 μS/cm. The breakthrough volumes from increasing concentration agreed with those from decreasing concentration. The volume required to achieve $C/C_0 = 0$, however, was 0.22 to 0.37 pore volumes larger than the volume required to achieve $C/C_0 = 1$.

The breakthrough of CHC occurred 0.03 to 0.16 pore volumes later than was the case for NaCl. With the exception of DCM, $C/C_0 = 1$ was reached 0.07 to 0.29 pore volumes later than for NaCl. When the concentration was decreasing, the first breakthrough of CHC-free water occurred 0.05 to 0.17 pore volumes later than did the breakthrough of NaCl-free water. The water was totally free of CHC after 1.25 to 1.47 pore volumes. The different PER concentrations in V1 and V2 did not seem to have any effects on the results.

The distribution coefficient K_d describes the relationship between the concentrations of the compound of interest in the solid and solution phases:

$$K_d = \frac{V}{m}\frac{M_S}{M_L} = \frac{C_S}{C_L} \text{ (mL/g)}$$

where V = volume of the solution
M_S = mass of sorbed solute
m = mass of porous medium
M_L = mass of the solute in the solution
C_S = concentration of sorbed solute
C_L = concentration of the solute in the solution

The K_d values obtained from the breakthrough experiments conducted here are shown in Table 7.

The retardation which a solute will exhibit relative to the velocity of an ideal nonsorbing tracer (and, therefore, also relative to the velocity of the water itself) will depend on K_d. The relationships are:

Table 7. K_d Values Obtained from Breakthrough Experiments

Experiment	K_d (mL/g)
V1 (PER, C_0 = 68.0 mg/L)	0.033
V2 (PER, C_0 = 11.2 mg/L)	0.045
V3 (TCE, C_0 = 52.5 mg/L)	0.008
V4 (DCM, C_0 = 80.0 mg/L)	0.011

$$\frac{u_{solution}}{u_{water}} = u_{relative} = \frac{1}{1 + K_d\gamma/n}$$

where $u_{solution}$ = mean velocity of the sorbing solute
u_{water} = velocity of the water
$u_{relative}$ = relative flow velocity
γ = bulk density of the porous medium
n = porosity.

If

$$K_d\gamma/n = D$$

and

$$R_D = \text{retardation factor} = (1 + D)$$

then

$$u_{relative} = 1/R_D$$

The BT values for the CHCs and NaCl yielded the results shown in Table 8.

Table 8. BT Values for CHC and NaCl from Breakthrough Experiments

Experiment	BT for CHC	BT for NaCl	R_D	$u_{relative}$
PER (V1, C_0 = 68.0 mg/L)	0.87	0.75	1.16	0.86
PER (V2, C_0 = 11.2 mg/L)	0.89	0.73	1.22	0.82
TCE (V3, C_0 = 52.5 mg/L)	0.84	0.81	1.04	0.96
DCM (V4, C_0 = 80.0 mg/L)	0.79	0.75	1.05	0.95

These laboratory results may not apply to the retardation of CHCs at concentrations which are very low (<1 mg/L). For the concentrations examined here, though, it may be concluded that these CHCs will move in natural ground-water at velocities which are only negligibly slower than that of the water itself. When the concentrations are high, such as will occur when directly in, or even near, the actual core of a spill, the transport rate should be taken to be the water velocity itself since there may be no retardation under those conditions.

8

Spreading as a Gas Phase

We have repeatedly emphasized the importance of the gas zones formed by the subsurface vaporization of chlorinated hydrocarbons (CHCs). We have developed experimental techniques to obtain a quantitative understanding of this process. The principles of the techniques will be discussed using the results of an experiment with chloroform as an example (Figures VIb and 41).

The porous medium for the experiment was a very uniform, coarse sand (Curve 3, Figure 1). The column was sealed at the top and bottom with reducers of the glass bell type. The pressure was equalized across the column with one vent on each of the two glass reducers. The vessel used to add the CHC was located on the upper reducer. A dish lined with filter paper was positioned on a wire frame on top of the sand surface. A narrow sampling port was installed on the bottom glass reducer; an injection needle was just able to fit through the port. The room temperature was 20 to 22°C.

Prior to the beginning of the experiment, air was pulled through the column with a pump. The column was thereby dried and cleaned of any residual CHC. At the beginning of the run, 31 mL of chloroform were added in drops onto the filter paper. The addition took place in four separate portions of 7 to 8 g and was complete in 50 min. The first three additions each vaporized within 5 min, and the fourth required 30 min. These vaporization times were obtained by observing the filter paper. From the beginning of the experi-

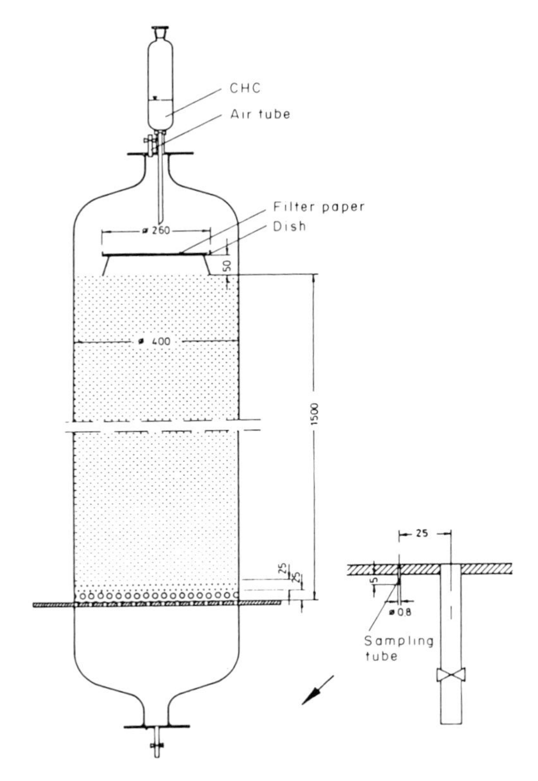

Figure 41. Experimental apparatus for the determination of the sinking of gaseous CHC in a porous medium.

ment on, the two vents were maintained in a half-open position to equalize the pressure.

After 2 hr and 20 min, chloroform was detected at a concentration of 32 μg/L in the gas in the lower glass reducer (Figure 42). A steep increase in the concentration occurred immediately thereafter. After a total of 6 hr and 30 min, a maximum concentration of 324 mg/L was reached. The decline in the concentration was somewhat less steep than the initial concentration increase. The distribution of the chloroform gas throughout the entire pore volume occurred surprisingly quickly. In this top to bottom mode, dichloro-

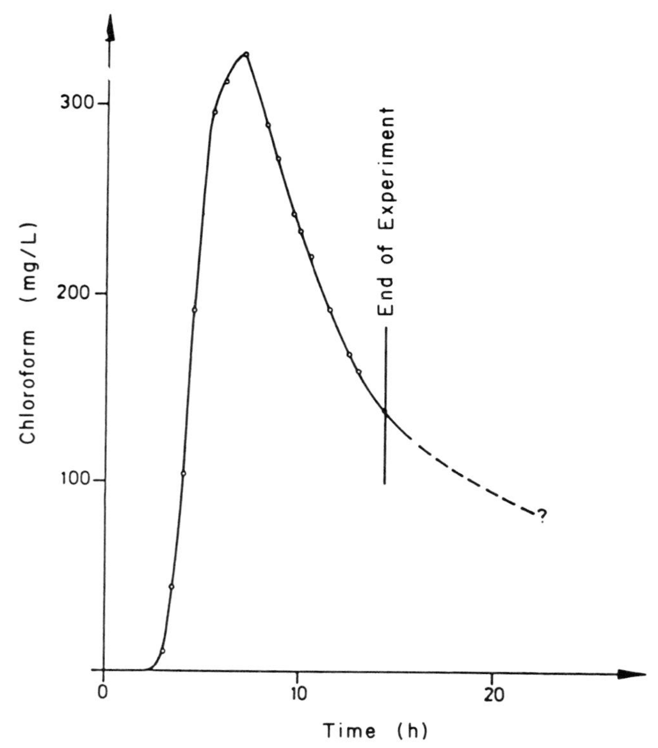

Figure 42. Time-dependent gas-phase chloroform concentration in the lower glass reducer of the apparatus presented in Figure 41.

methane (DCM) was found to move through the porous medium at a rate similar to chloroform; 1,1,1-trichloroethane (1,1,1-TCA) moved at a somewhat slower rate.

Since the density of air saturated with a CHC will increase as the CHC vapor pressure increases, the latter will affect gas phase spreading because gravity will act differently on gas masses of different densities. The importance of this effect was shown by an experiment with DCM moving in the column *from the bottom toward the top*. One hundred milliliters of DCM were added to a dish in the lower glass reducer. Now, 7 hr were required for measurable amounts of DCM to be found in the upper reducer. Thereafter, the concentration climbed slowly to a maximum of 600 mg/L over 5 to 6 days.

While the two DCM experiments cannot be compared in all respects, they do indicate that vaporization in the unsaturated zone is indeed important. After vaporization occurs from a spill, the dense CHC gases will be affected by gravity and will tend to sink. For example, the relative vapor densities that are tabulated in some handbooks for the CHC compounds range from 2.9 to 5.7.

[Extended Translator's Note: Dr. Schwille's "relative vapor densities" are the ratios of the *hypothetical* densities of pure CHC vapor relative to the density of dry air at the same temperature and pressure. The densities are *hypothetical* inasmuch as none of the CHC compounds of interest have a pure compound vapor pressure of 1 atm. For each compound, then, a more realistic way to look at this issue would be to compute the ratio of the density of dry air *saturated* with the CHC of interest at 293 K and 1 atm total pressure relative to pure dry air (which has an average molecular weight of 29.0) at the same temperature and pressure. See Appendix I (Translator's Appendix) for a detailed explanation of the mathematical formula for determining these true relative vapor density (RVD) values. When this formula is used, the true RVDs at 20°C are determined to be as shown in Table 9.

Table 9. True Relative Vapor Densities at 20°C

Compound	Molecular Weight (g)	Vapor Pressure (atm) (20°C)	Relative Vapor Density
Dichloromethane (DCM)	84.9	0.46	1.89
Chloroform	119.4	0.20	1.62
Trichloroethylene (TCE)	131.4	0.076	1.27
1,1,1-Trichloroethane (1,1,1-TCA)	133.4	0.13	1.47
Tetrachloroethylene (PER)	165.8	0.018	1.09

(End of Translator's Note.)]

In systems in which lateral spreading is not restricted, the spreading will occur as in Figure IIIa. The issue of the diffusion of gases through the porous medium and upward toward the ground surface has been investigated by the Institute for Water, Soil, and Air Hygiene of the Federal Health Office. For this topic, the reader is referred to their manual on CHCs.

9

Conclusions

We believe that we have succeeded in using experiments with physical models to lay out the principles governing the spreading of chlorinated hydrocarbons (CHCs) in the subsurface environment in a plausible manner. The results have validated our initial working hypotheses. Now that the general scenarios for the spreading of CHCs have been worked out, it is absolutely necessary that extended series of experiments be carried out which determine the individual parameters that control spreading. This should at least be done for the most important soil types and for several pertinent soil water contents. Since the experimental results which we have obtained to date pertain, strictly speaking, only to the conditions under which they were obtained, one should be on guard against using observed generalizations of behavior in an excessively broad manner.

For example, the term *sinker* (in contrast to *floater*) is often used in an overly simplistic way to refer to the fact that CHC compounds are relatively dense. This type of perception leads to the conclusion that a given CHC will always sink in an aquifer system whenever there is enough CHC to exceed residual saturation. In fact, we now know that the sinking of CHCs in a water-saturated medium requires a certain CHC pressure in addition to excess CHC. That required pressure increases as the pore size of the medium decreases; in the smaller pore size media, the downward movement of a CHC usually comes to a halt quickly.

Another example is provided by the fact that CHC compounds are generally classified as very volatile. This is indeed the case. However, this ignores the fact that the CHCs exhibit a range of volatilities; under practical spill situations, this range can cause different behaviors. Thus, in a dry medium, a given amount of dichloromethane (DCM) will require 12 hr to become 95% volatilized and sink (as a gas) to a depth of 1.5 m. For the same to occur with perchloroethylene (tetrachloroethylene, or PER) would require 12 days. For DCM, then, if a spill occurred slowly, the complete mass of DCM could volatilize in the unsaturated zone. Compared to DCM, PER volatilizes slowly; under the same spill conditions, liquid PER might sink down to and into the saturated zone.

As another example, few spill studies determine the water content of the porous medium. There is a big difference in behavior, however, when a solvent is spilled under a building where the soil is dry vs in the open where the soil is moist. In a dry medium, a given CHC can occupy the corners of the pores. It can also be held in substantial quantities in a medium with small pores. In a wet medium, however, these spaces are occupied by water. Under certain conditions, then, a spilled CHC may infiltrate more rapidly in a moist medium than in a dry medium since all of the porosity is not available in the former to retain the CHC in residual saturation.

It has been claimed repeatedly that concrete is not suitable for impoundments since there is no concrete that is completely leak-free with respect to CHC. The real-world examples of leaks that have been studied do not permit one to conclude whether CHC is leaking (1) as a fluid phase through cracks or hairline fractures; (2) as a solute in water; and/or (3) as a gas through porous portions of the concrete. Systematic laboratory experiments are needed.

Details such as these play an important role in determining what measures can be taken to prevent as well as remedy contamination. However, they are perceived quite differently by those who are responsible for contamination and those who are damaged by it. It is not unusual for excess remediation costs to be caused by the uncertainties of the practitioners of this field. The examples discussed in this book may have indeed demonstrated that when it comes to the topic of behavior of CHCs in the subsurface, many specialized questions are in need of attention. The pursuit of the answers to those questions is justified on the basis of economic reasons alone. Although physical models are sometimes viewed as outmoded and archaic, they are capable of providing the correct conceptualizations of the CHC spreading processes. In addition, in the event that the appropriate mathematical models should one day be developed, the results will provide suitable test data.

The use of models in experiments does, to be sure, have its limitations. The large number of variables which can be important during a spill in the natural environment cannot be taken into full consideration in the laboratory. It will therefore be absolutely necessary to carefully collect the facts and results obtained in the remediation of spills and to extract whatever information possible from them. Those data should also be compared with results obtained in model experiments.

Appendix I

Translator's Appendix: Physical and Chemical Properties of Dense Solvent Compounds*

*It is extremely important to note that a model that describes the simultaneous processes of free product vaporization, free product solubilization, and air/water partitioning must use a consistent set of the parameters p_o, S, and H; however, the values of these parameters in this table are not necessarily consistent, because they were obtained by different investigators.

Table 1. Physical and Chemical Properties of Dense Solvent Compounds

Compound	MW (g)	S (mg/L)	P_o (torr)	K_{oc} (mL/g)	d (g/cm³)	BP (°C)	μ Absolute Viscosity (cp)	ν Kinematic Viscosity	Relative H (atm-m³/mol)	Vapor Density[a]
Non-Aromatics										
dichloromethane (DCM)	84.9	20000	349	8.8	1.33	40	0.44	0.32	0.0017	1.89
chloroform	119.4	8200	151	44	1.49	62	0.56	0.38	0.0028	1.62
bromodichloromethane	163.8	4500	50	61	1.97	90	1.71	0.87	0.0024	1.31
dibromochloromethane	208.3	4000	76	84	2.38	119	–	–	0.00099	1.62
bromoform	252.8	3010	5	116	2.89	150	2.07	0.72	0.00056	1.05
trichlorofluoromethane	137.4	1100	667	159	1.49	24	–	–	0.11	4.28
carbon tetrachloride	153.8	785	90	439	1.59	77	0.97	0.61	0.023	1.51
1,1-dichloroethane	99.0	5500	180	30	1.17	57	0.50	0.43	0.0043	1.57
1,2-dichloroethane	99.0	8690	61	14	1.26	83	0.84	0.67	0.00091	1.19
1,1,1-trichloroethane (1,1,1-TCA)	133.4	720[b]	100	152	1.35	74	0.84	0.62	0.013	1.47
1,1,2-trichloroethane	133.4	4500	19	56	1.44	114	–	–	0.00074	1.09
1,1,2,2-tetrachloroethane	167.9	2900	5	118	1.60	146	1.76	1.12	0.00038	1.03
1,1-dichloroethylene	97.0	400	590	65	1.22	32	0.36	0.30	0.021	2.54
1,2-dibromoethane (EDB)	187.9	4	11[b]	–	2.18	132	1.72	0.79	0.00082	1.08
1,2-cis-dichloroethylene	97.0	800[b]	200[b]	–	1.28	60	0.48	0.38	0.0029	1.62
1,2-trans-dichloroethylene	97.0	600	326	59	1.26	48	0.40	0.32	0.072	2.01
trichloroethylene (TCE)	131.5	1100	58	126	1.46	87	0.57	0.39	0.0071	1.27
tetrachloroethylene (PER)	165.8	200	14	364	1.63	121	0.90	0.54	0.0131	1.09
1,2-dichloropropane	113.0	2700	42	51	1.16	97	–	–	0.0023	1.16
trans-1,3-dichloropropylene	110.0	1000	25	48	1.22	112	–	–	0.0013	1.09

Ethers										
bis(chloromethyl) ether	115.0	22000	30[b]	1.2	1.32	104	–	–	0.00021	1.12
bis(2-chloroethyl) ether	143.0	10200	0.7	14	1.22	178	2.41	1.98	0.000013	1.004
bis(2-chloroisopropyl) ether	171.1	1700	0.9	61	1.11	187	–	–	0.00011	1.005
2-chloroethyl vinyl ether	106.6	15000	27	6.6	1.05	108	–	–	0.00025	1.10
Monocyclic Aromatics										
chlorobenzene	112.6	488[b]	12	330	1.11	132	0.80	0.72	0.0036	1.05
o-dichlorobenzene	147.0	100	1.0	1700	1.31	180	1.41	1.28	0.0019	1.005
m-dichlorobenzene	147.0	123[b]	2.3[b]	1700	1.29	172	1.08	0.84	0.0036	1.01

Source: See below for a list of references used to compile the data in this table.

Note: Temperature of measurement is 20°C unless otherwise noted. MW = molecular weight (g); S = solubility in water (mg/L or ppm); p_o = vapor pressure (torr or mm Hg); K_{oc} = sediment/water partition coefficient (mL/g); d = density (g/cm^3); BP = boiling point at 760 torr pressure (°C); μ = absolute viscosity (centipoise); ν = kinematic viscosity; H = Henry's Law constant for partitioning between air and water (atm-m^3/mole); and RVD = vapor density relative to dry air (dimensionless). See footnote *a* below on method used to calculate RVD values. Note that an estimate of H may be obtained from the values of p_o and S. However, no effort was made to make that value of H consistent with the tabulated value given below, since the p_o, S, and H values were all determined independently, and there was no way to give greater weight to any two out of three of the measurements, which would have been necessary to obtain a consistent set of the three parameters.

[a]RVD values have been calculated as the density of dry air saturated with the compound of interest at 20°C. It represents the weighted mean molecular weight of the compound-saturated air relative to the mean molecular weight of dry air which is 29.0 g/mol. If MW = molecular weight of the compound of interest, the RVD value may be calculated as:

$$RVD = \frac{\frac{p_o}{760} MW + \frac{760 - p_o}{760} 29.0}{29.0}$$

If the RVD relative to air saturated with water is desired, then the mean molecular weight for moist air at 20°C (28.75) should be used in place of 29.0 in the above formula.

[b]Value measured at 25°C.

REFERENCES

Dean, J.A., Ed. *Lange's Handbook of Chemistry*, (NY: McGraw-Hill, 1979).

Flick, E.W. "Industrial Solvents Handbook," Noyes Data Corporation, Park Ridge, NJ (1985).

Gosset, J.M. "Measurement of Henry's Law Constants for C_1 and C_2 Chlorinated Hydrocarbons," *Environ. Sci. Technol.* 21:202–208 (1987).

Mabey, W.R., et al. "Aquatic Fate Processes Data for Organic Priority Pollutants," U.S. EPA Report No. 440/4–81–014 (December 1982).

Riddick, J.A., and W.B. Bunger. *Organic Solvents, Techniques of Chemistry, Vol. II,* 3d ed. (NY: Wiley-Interscience, 1970).

Sedivec, V., and J. Flek. *Handbook of Analysis of Organic Solvents* (NY: John Wiley and Sons, Inc., 1976).

Timmermans, J. *Physico-Chemical Constants of Pure Organic Compounds* (NY: IUPAC, Elsevier Publishing, 1950).

Versheueren, K. *Handbook of Environmental Data on Organic Chemicals,* 2d ed. (NY: Van Nostrand Reinhold, 1983).

Weast, R.C., Ed. *Handbook of Chemistry and Physics* (Boca Raton, FL: CRC Press, Inc.).

Bibliography

It was not possible within the scope of this report to give a complete overview of the literature on this topic. The reader who desires such information is referred to the well-known "Leitfaden für die Beurteilung und Behandlung von Grundwasserverunreinigung durch leichtflüchtige Chlorkohlenwasserstoffe" [Guide for the Assessment and Remediation of Groundwater Contamination by Volatile Chlorinated Hydrocarbons]. This guide was published by the Ministry for Nutrition, Agriculture, Environment, and Forestry/Baden-Württemberg. It surveys all of the most important literature references, as well as the reports mentioned in this book from the institutes participating in the publication of the guide.

[Translator's Note: For convenience, the bibliography of the aforementioned report has been reproduced in this book. The titles of the German articles and reports are given in both German and English.]

Ahrens, G., 1983: Auswaschen von chlorierten Kohlenwasserstoffen [Removal of chlorinated hydrocarbons]. Gutachten an der Landesanstalt für Umweltschutz [Expert opinion of the regional office for environmental protection], Karlsruhe (unveröffentlicht) [unpublished].

Assman, W., et al., 1983: Tiefe Gw-Messstellen im Lockergestein, Erfahrungen und Weiterentwicklung [Deep

Source: Reprinted from "Leitfaden für die Beurteilung und Behandlung von Grundwasserverunreinigung durch leichtflüchtige Chlorkohlenwasserstoffe" [Guide for the Assessment and Remediation of Groundwater Contamination by Volatile Chlorinated Hydrocarbons].

groundwater monitoring points in porous rock: results and further developments]. bbr, Heft [No.] 2, 34 [vol. 34].

Aurand, K., Fischer, M., 1981: Gefährdung von Grund- und Trinkwasser durch leichtflüchtige Chlorkohlenwasserstoffe [Threats to groundwater and drinking water from volatile chlorinated hydrocarbons]. *WaBoLu-Berichte* [Water-Soil-Air Report] 3/1981, Dietrich Reimer Verlag, Berlin.

Aurand, K., Friesel, P., Milde, G., Neumayr, V., 1981: Behavior of Organic Solvents in the Environment. In: *Quality of Groundwater, Proceedings of an International Symposium, Noordwijkerhout, The Netherlands*. Studies in Environmental Science, Volume 17, Elsevier Publishing Co., The Netherlands.

Baldauf, G., 1983: Der Fall Grenzach–Beispiel einer Grundwasserverschmutzung mit umweltrelevanten Stoffen [The Grenzach accident–an example of groundwater contamination with environmentally relevant substances]. *DVGW-Schriftenreihe Wasser* [DVGW-Water Series], Heft [No.] 29, 53–69.

Bauer, U., 1981: Belastung des Menschen durch Schadstoffe in der Umwelt–Untersuchungen über leichtflüchtige, organische Halogenverbindungen in Wasser, Luft, Lebensmitteln und im menschlichen Gewebe [The burdening of humans by toxic substances in the environment–investigations on volatile organic halogenated compounds in water, air, food, and fabrics of apparel]. *Zbl. Bakt. Hyg.*, I. Abt. Orig. B 174.

Bear, J., 1972: *Dynamics of Fluids in Porous Media*. American Elsevier Publishing Co., Inc., New York.

Bear, J., Jacobs, M., 1965: On the movement of waterbodies injected into aquifers. *J. Hydrol.*, Vol. 3, 37–57.

Bertsch, W., 1978: Die Koeffizienten der longitudinalen und transversalen hydrodynamischen Dispersion–ein Literaturüberblick [The coefficients of longitudinal and transverse hydrodynamic dispersion–a literature overview]. Deutsche Gewässerkundliche Mitteilungen [German

Hydrological Communications], *DGM* [J. Hydrol. Dept. of the Republic and the Regions] 22, (2), 37–46.

Bolt, H. M., 1980: Die toxikologische Beurteilung Halogenierter Äthylene [The toxicological evaluation of chlorinated ethylene]. *Arbeitsmedizin/Sozialmedizin/Präventivmedizin* [Occupational Health, Social Medicine, and Preventive Medicine], Heft [No.] 3.

Bolt, H. M., 1981: Die toxikologische Beurteilung einiger halogenierter Kohlenwasserstoffe [The toxicological evaluation of selected chlorinated hydrocarbons]. In: Aurand, K.; Fischer, M. [Eds]; Gefährdung von Grund- und Trinkwasser durch leichtflüchtige Chlorkohlenwasserstoffe [Threats to groundwater and drinking water from volatile chlorinated hydrocarbons]. *WaBoLu-Berichte* [Water-Soil-Air Report] 3/1981, 46–54, Dietrich Reimer Verlag, Berlin.

Briggs, G.G., 1981: Theoretical and experimental relationships between soil adsorption, octanol-water partition coefficients, water solubilities, bioconcentration factors, and the parachor. *J. Agric. Food Chem.*, Vol. 29, 1050–1059.

Brunner, W., Straub, D., Leisinger, T., 1980: Bacterial degradation of dichloromethane. *Appl. Environ. Microbiol.*, Vol. 40, 950–958.

Bundesgesundheitsamt [German Federal Health Office], 1982: Empfehlung zum Vorkommen von flüchtigen Halogenkohlenwasserstoffen im Grund- und Trinkwasser [Recommendations concerning the occurrence of volatile halogenated hydrocarbons in groundwater and drinking water]. *Bundesgesundh*. Bl. 25, Nr. 3 März [National Health 25 (3) March].

Bundesverband der Unfallversicherungsträger der öffentlichen Hand e. V., München [German Federal Association of Publicly Supported Accident Insurers, Munich] (Herausgeber), 1978a: Sicherheitsregeln für den Umgang mit aliphatischen Chlorkohlenwasserstoffen und deren Gemischen – CKW-Regeln für den Betrieb [Safety Regulations for the Handling of Chlorinated Aliphatic Hydrocarbons and their Mixtures – Chlorinated Hydrocarbon Rules for the Factory]. Ausgabe September.

Bundesverband der Unfallversicherungsträger der öffentlichen Hand e. V., München [German Federal Association of Publicly Supported Accident Insurers, Munich] (Herausgeber), 1978b: Merkblatt für den Umgang mit Chlorkohlenwasserstoffen – CKW-Merkblatt für den Beschäftigten [Pamphlet on the Handling of Chlorinated Hydrocarbons – Chlorinated Hydrocarbon Pamphlet for Employees]. Ausgabe September.

Bundesverband der Unfallversicherungsträger der öffentlichen Hand e. V., München [German Federal Association of Publicly Supported Accident Insurers, Munich] (Herausgeber), 1978c: Merkblatt für den Umgang mit lösemittelhaltigen Arbeitsstoffen zur Kaltreinigung (Kaltreiniger-Merkblatt) [Pamphlet on the Handling of Materials Containing Solvents for use in Dry Cleaning (Dry Cleaning Pamphlet)]. Ausgabe September.

Bundesverband der Unfallversicherungsträger der öffentlichen Hand e. V., München [German Federal Association of Publicly Supported Accident Insurers, Munich] (Herausgeber), 1978d: Merkblatt für den Umgang mit Lösemitteln sowie brennbaren organischen Flussigkeiten, Kohlenwasserstoffe, Halogenkohlenwasserstoffe, Alkohole, Ather, Ketone u.a. z. B. Schwefelkohlenstoff [Pamphlet on the Handling of Solvents such as Flammable Organic Fluids, Hydrocarbons, Halogenated Hydrocarbons, Alcohols, Ethers, Ketones, and Others, for example, Carbon Disulfide]. Ausgabe Oktober.

Chiou, C. T., Freed, V. H., Schmedding, D. W., Kohnert, R. L., 1977: Partition coefficient and bioaccumulation of selected organic chemicals. *Environ. Sci. Technol.*, Vol. 11, 475–478.

Chiou, C. T., Peters, L. J., Freed, V. H., 1979: A physical concept of soil-water equilibria for nonionic organic compounds, *Science*, Vol. 206, 831–832.

Chiou, C. T., Schmedding, D. W., 1982: Partitioning of

organic compounds in octanol-water systems. *Environ. Sci. Technol.*, Vol. 16, 4–10.

Conrads, H., 1981: Chlorkohlenwasserstoffprobleme der Stadtwerke Heidelberg AG [Chlorinated hydrocarbon problems of Heidelberg City Works, Inc.]. In: Aurand, K.; Fischer, M. [Eds.]: Gefährdung von Grund- und Trinkwasser durch leichtflüchtige Chlorkohlenwasserstoffe [Threats to groundwater and drinking water from volatile chlorinated hydrocarbons]. *WaBoLu-Berichte* [Water-Soil-Air Report] 3/1981, 73–74, Dietrich Reimer Verlag, Berlin.

Dietzel, F., 1981: Schadensermittlung, Verursachersuche, und Sanierungsmassnahmen [Report on damages, the search for the responsible party, and remediation measures]. In: Gefährdung von Grund- und Trinkwasser durch leichtflüchtige Chlorkohlenwasserstoffe [In: Aurand, K.; Fischer, M. (Eds.): Threats to groundwater and drinking water from volatile chlorinated hydrocarbons]. *WaBoLu-Berichte* [Water-Soil-Air Report] 3/1981, 67–69, Dietrich Reimer Verlag, Berlin.

Esseger, W., 1983: Geologische Aspekte bei der Behandlung von HKW-Schadensfällen [Geological aspects concerning handling accidents involving halogenated hydrocarbons]. *DVGW-Schriftenreihe Wasser* [DVGW-Water Series], Heft [No.] 36, (im Druck) [in press].

Fricke, H., 1981: Grundwasserverunreinigung durch Vorstellung und Anwendung von chlorierten Lösungsmitteln aus der Sicht der Industrie [The perspective of industry on groundwater contamination due to the production and use of chlorinated solvents]. In: *DVGW Schriftenwerke*, Heft [No.] 29.

Friesel, P., Neumayr, V., Milde, G., 1980, 1981, 1982, 1983: Beeinflussung der Grundwasserqualität durch Umweltchemikalien im Boden [The influence on groundwater quality of environmental chemicals in the subsurface]. Zwischenberichte und Abschlussbericht zum Forschungsvorhaben: Methoden zur ökotoxikologischen Bewertung von Chemikalien [Interim reports and final report for the research project: Methods for the ecotoxicological evaluation of chemicals]. Institut für Wasser-, Boden- und Lufthygiene des Bundesgesundheitsamtes [Institute for Water, Soil, and Air Hygiene of the Federal Health Office], Berlin (unveröffentlicht) [unpublished].

Fritschi, G., Neumayr, V., Schinz, V., 1979: Tetrachlorethylen und Trichlorethylen im Trink- und Grundwasser [Tetrachloroethylene and trichloroethylene in drinking and groundwater]. *WaBoLu-Berichte* [Water-Soil-Air Report], 1/1979, Dietrich Reimer Verlag, Berlin.

Fujita, T., Iwasa, J., Hansch, C., 1964: A new substituent constant, derived from partition coefficients. *J. Am. Chem. Soc.* Vol. 86, 5175–5180.

Hamaker, J. W., Thompson, J. M., 1972: In: *Organic Chemicals in the Soil Environment*. M. Dekker, New York.

Hansch, C., Leo, A., 1979: *Substitute Constants for Correlation Analysis in Chemistry and Biology*. John Wiley and Sons, Inc., NY.

Henschler, D., 1980: Carcinogenicity study of trichlorethylene by long-term inhalation in three animal species. *Arch. Toxicol.* Vol. 43, 237–248.

Hommel, 1980. *Handbuch der gefährlichen Güter* [Handbook of Dangerous Substances], Springer-Verlag, Berlin.

Irmer, H., Knüpfer, A., Neumayr, V., 1981: Studie über Stickstoff- und Metalleinleitungen sowie Einleitungen von halogenierten Kohlenwasserstoffen in das Abwasserkanalnetz der Stadt Langen, Studie im Auftrag der Stadtwerke Langen [Studies on the Inputs of Nitrogen, Metals, and Halogenated Hydrocarbons to the Wastewater System of the City of Langen: Studies Commissioned by the City Works of Langen]. Institut für Wasser-, Boden- und Lufthygiene des Bundesgesundheitsamtes [Institute for Water, Soil, and Air Hygiene of the Federal Health Office], Berlin], (unveröffentlicht) [unpublished].

Karickhoff, S. W., Brown, D. S., Scott, T. A., 1979: Sorption of hydrophobic pollutants on natural sediments. *Water Res.*, Vol. 13, 241–248.

Kauch, E. P., 1982: Zur Situierung von Brunnen im

Grundwasserstrom [On the siting of wells in systems of moving groundwater]. *Östereichische Wasserwirtschaft* [Austrian Water Resources], Heft [No.] 7/8, 157–162.

Kinzelbach, W., 1983: Sanierungsmassnahmen im Vorfeld von Trinkwasserfassungen [Remedial measures carried out in areas surrounding drinking water well fields]. In: Vermeidung und Sanierung von Grundwasserverunreinigungen [Avoidance and remediation of groundwater contamination], *DVGW-Schriftenreihe Wasser* [DVGW-Water Series], Heft [No.] 36 (im Druck) [in press].

Klotz, D., 1973: Untersuchungen zur Dispersion in pörosen Medien [Investigations of dispersion in porous media], *Z. Dentsch Geol. Ges.* [J. German Geological Society], Vol. 124, 523–533.

Kobus, H., 1981: Strömungsmechanische Grundlagen des Transports der Halogenkohlenwasserstoffe im Grundwasserleiter—Massnahmen zur Erfassung des verunreinigten Wassers [The fluid mechanical basis of transport of halogenated hydrocarbons in a groundwater aquifer—measures for the capture of contaminated waters]. In: Halogenkohlenwasserstoffe in Grundwässern [Halogenated hydrocarbons in groundwaters], *DVGW-Schriftenreihe Wasser* [DVGW-Water Series], Nr. [No.] 29, 91–103.

Konikow, L. F., Bredehoeft, T. D., 1978: Computer Model of Two-Dimensional Solute Transport and Dispersion in Groundwater. Techniques of Water Resources Investigations of the United States Geological Survey, Book 7, Chapter C2, United States Government Printing Office, Washington, DC.

Lambert, S. M., 1968: Omega, a useful index of soil sorption equilibria. *J. Agric. Food Chem.*, Vol. 16, No. 2.

Länderarbeitsgemeinschaft Abfall [Regions' Working Group on Wastes] (LAGA): Informationschrift Gefährdungsabschätzung und Sanierungsmöglichkeiten bei Altablagerungen [Information Pamphlet: Hazard Estimation and Remedial Cleanup Possibilities at Defunct Waste Disposal Sites]. Erich-Schmidt Publishing Co., Bielefeld.

Landesanstalt für Umweltschutz Baden-Württemberg [Regional Office for Environmental Protection], 1981: Institut für Wasser- und Abfallwirtschaft: Ganglinien der Grundwasserstände im Dienstbezirk des WWA Heidelberg [Institute for Water and Wastewater Management: Trends in the Groundwater Levels in the Service Area of Heidelberg], 1961–1980.

Lenda, A., Zuber, A., 1970: Tracer Dispersion in Groundwater Experiments. In: *Isotope Hydrology*, IAEA-SM-129/37, 619–641.

Leo, A., Hansch, C., Elkins, D., 1971: Partition coefficients and their uses. *Chem. Rev.*, Vol. 71, No. 6.

Löchner,F., 1981: Lösungsmittel aus leichtflüchtigen Halogenkohlenwasserstoffen (HKW) [Solvents from volatile halogenated hydrocarbons]. Produktion, Einsatzbereiche und Verwendungstechnologien [Production, end uses, and technologies for use]. In: Aurand, K.; Fischer, M. [Eds.]; Gefährdung von Grund- und Trinkwasser durch leichtflüchtige Chlorkohlenwasserstoffe [Threats to groundwater and drinking water from volatile chlorinated hydrocarbons]. *WaBoLu-Berichte* [Water-Soil-Air Report] 3/1981, Dietrich Reimer Verlag, Berlin.

Ministerium für Ernährung, Landwirtschaft, Umwelt und Forsten, Baden-Württemberg (MELUF) [Ministry for Nutrition, Agriculture, Environment, and Forestry/Baden-Württemberg] (Herausgeber), 1980: Hydrogeologische Kartierung und Grundwasserbewirtschaftung Rhein-Neckar-Raum, Analyse des 1st-Zustandes [Hydrogeological Mapping and Groundwater Management of the Rhein/Neckar Region, Analysis of Phase 1].

Ministerium für Ernährung, Landwirtschaft, Umwelt und Forsten, Baden-Württemberg (MELUF) [Ministry for Nutrition, Agriculture, Environment, and Forestry/Baden-Württemberg] (Herausgeber), 1983: Informationsschrift zum Umgang mit chlorierten Kohlenwasserstoffen [Information Pamphlet on the Handling of Chlorinated Hydrocarbons].

Ministerium für Arbeit, Gesundheit und Sozialordnung

Baden-Württemberg [Ministry for Labor, Health, and Social Programs/Baden-Württemberg], 1981. Erlass vom 3.11. 1981 [Released 3/11/1981] (VI/6–8744.10/81).

Nagel, G., 1981: Belüftungsverfahren in der Trinkwasseraufbereitung [Aeration in pretreating drinking water]. Institut für Bauingenieurwesen V der TU München: Berichte aus Wassergütewirtschaft und Gesundheitswesen [Institute for Construction Engineering, V/TU, Munich: Reports concerning water quality management and public health], Nr. [No.] 36, 41–70.

Neely, W. B., 1979: Estimating rate constants for the uptake and clearance of chemicals by fish. *Environ. Sci. Technol.*, Vol. 13, 1506–1410.

Nelson, R., 1978: Evaluating the environmental consequences of groundwater contamination. *Water Resour. Res.*, Vol. 14, 409–450.

Neumayr, V., 1981a: Beiträge zur Entwicklung Systematischer Sanierungskonzepte bei Boden- und Grundwasserkontaminatione [Contribution for the development of systematic concepts for the remediation of soil and groundwater contamination]. In: Aurand, K. [Ed.]: *Bewertung chemischer Stoffe im Wasserkreislaut* [Evaluation of chemicals in the hydrological cycle], 37–52, Erich-Schmidt Verlag, Berlin.

Neumayr, V., 1981b: Verteilungs- und Transportmechanismen von chlorierten Kohlenwasserstoffen in der Umwelt [Distribution and transport mechanisms of chlorinated hydrocarbons in the environment]. In: Aurand, K.; Fischer, M. [Eds.]; Gefährdung von Grund- und Trinkwasser durch leichtflüchtige Chlorkohlenwasserstoffe [Threats to groundwater and drinking water from volatile chlorinated hydrocarbons]. *WaBoLu-Berichte* [Water-Soil-Air Report] 3/1981, 24–38, Dietrich Reimer Verlag, Berlin.

Neumayr, V., 1982: Studie über die Einleitung halogenierten Kohlenwasserstoffen in das Abwasserkanalnetz der Stadt Langen, Institut für Wasser-, Boden- und Lufthygiene des Bundesgesundheitsamtes [Study on the Input of Halogenated Hydrocarbons to the Wastewater System of the City of Langen, Institute for Water, Soil, and Air Hygiene of the Federal Health Office] (unveröffentlicht) [unpublished].

Neumayr, V., 1983a: Zweite Teilstudie über die Einleitung halogenierter Kohlenwasserstoffe in das Abwasserkanalnetz der Gemeinde Egelsbach, Institut für Wasser-, Boden- und Lufthygiene des Bundesgesundheitsamtes [Study on the Input of Halogenated Hydrocarbons to the Wastewater System of the Municipality of Egelsbach, part 2, Institute for Water, Soil,and Air Hygiene of the Federal Health Office] (unveröffentlicht) [unpublished].

Neumayr, V., 1983b: Möglichkeiten und Grenzen der Erfassung von Untergrundverunreinigungen durch halogenierte Kohlenwasserstoffe, [Possibilities and limits for the capture of subsurface contamination by halogenated hydrocarbons]. *DVGW-Schriftenreihe Wasser* [DVGW-Water Series], Heft [No.] 36 (im Druck) [in press].

Pickens, F. J., Grisak, E. G., 1980: Scale-dependent dispersion in a stratified granular aquifer. *Water Resour. Res.*, Vol. 17, 1191–1211.

Piet, G. J., Zoetemann, B. C. J., 1980: Organic water quality changes during sand bank and dune filtration of surface waters in the Netherlands. *J. Am. Water Works Assoc.*, Vol. 72, 400–404.

Roberts, P. V., Reinhard, M., Valocchi, A. J., 1982: Movement of organic contaminants in groundwater: Implications of Water Supply. *J. Am. Water Works Assoc.*, Vol. 150, 408–413.

Roberts, P. V., Schreiner, J., Hopkins, G. D., 1982: Field study of organic quality changes during groundwater recharge in the Palo Alto baylands. *Water Res.*, Vol. 16, 1025–1035.

Scheidegger, A. E., 1957: On the Theory of Flow Phases in Porous Media. Proc. IUGG General Assembly, Toronto, 2, 236–242.

Schwarzenbach, R. P., Westall, J., 1981: Transport of nonpolar organic compounds from surface water to groundwater. *Environ. Sci. Technol.*, Vol. 15, 1360–1367.

Schwarzenbach, R. P., 1982: Groundwater Pollution by

Hydrophobic Organic Compounds. Presented at the Course on Pollution and Quality Control of Groundwater, March 8–12, Zürich.

Schwille, F., 1981: Groundwater pollution in porous media by fluids immiscible with water. *The Science of the Total Environment*, Vol. 21, 173–185.

Schwille, F., 1982: Die Ausbreitung von Chlorkohlenwasserstoffen im Untergrund, erläutert anhand von Modellversuchen [The spreading of chlorinated hydrocarbons in the subsurface as illustrated with model experiments], *DVGW-Schriftenreihe Wasser* [DVGW-Water Series], Nr. [No.] 31, Eschborn, 203–234.

Schwille, F., 1983: The behaviour of organic fluids immiscible with water in the unsaturated zone. In: *Pollutants in the Unsaturated Zone,* B. Yaron, G. Dagan, and J. Goldschmid (Eds.), Springer-Verlag, Heidelberg and New York.

Sontheimer, H., 1983a: Verfahrenstechnische Grundlagen für Anlagen zur Entfernung von Halogenkohlenwasserstoffen aus Grundwässern [Process Engineering Rules for Facilities Involved in the Removal of Halogenated Hydrocarbons from Groundwater]. Vortragsreihe und Erfahrungsaustausch über spezielle Fragen der Wassertechnologie 1982 [Lecture series and information exchange concerning special water technology issues for 1982] in Karlsruhe und Mülheim/Ruhr, Bereich Wasserchemie [Department of Water Chemistry] am Engler-Bunte-Institut der Universität Karlsruhe, Heft [No.] 21.

Sontheimer, H., 1983b: Aufbereitung von verunreinigten Grundwässern im Wasserwerk [Treatment of contaminated groundwaters in water works]. *DVGW-Schriftenreihe Wasser* [DVGW-Water Series], Heft [No.] 36 (im Druck) [in press].

Sontheimer, H., Cornel, P., 1981: Physikalische-chemische Vorgänge beim Transport von Halogenkohlenwasserstoffen im Grundwasser [Physical-chemical processes for the transport of halogenated hydrocarbons in groundwater]. *DVGW-Schriftenreihe Wasser* [DVGW-Water Series], Heft [No.] 29.

Stucki, G. et al., 1981: Microbial degradation of chlorinated C_1 and C_2 hydrocarbons. In: Leisinger, T. et al. (Eds.): *Microbial Degradation of Xenobiotics and Recalcitrant Compounds (FEMS Symposium No. 12)*, Academic Press Inc., London, 131–137.

Trenel, J., Wolf, G., Neumayr, V., 1983: Organohalogenverbindungen im Wasserkreislauf–Analysenverfahren, Probenahmen und Bewertungen im Abwasserbereich und im Vorfluter. [Halogenated organic compounds in the hydrological cycle– analytical techniques, sample acquisition, and evaluation in waste waters and drainage channels]. Abschlussbericht für den Umweltforschungsplan des Bundesministers des Innern– Wasserwirtschaft, im Auftrag des Umweltbundesamtes [Final report for the environmental research plan of the Federal Ministry of the Interior for Water Resources, as commissioned by the Federal Environmental Office]. Institute für Wasser-, Boden-, und Lufthygiene des Bundesgesundheitsamtes [Institute for Water, Soil, and Air Hygiene of the Federal Health Office] (in Vorbereitung) [in preparation].

Veith, G. D., Austin, N. M., Morris, R. T., 1979: A rapid method for estimating log P for organic chemicals. *Water Res.*, Vol. 13, 43–47.

Werner, J., 1982: Erfahrungen bei Schadensfällen mit Chlorkohlenwasserstoffen [Experiences with accidents involving chlorinated hydrocarbons]. *Mit. Ing.-U. Hydrogeol.* [Commun. Engineering and Hydrogeol.], Vol. 13, 131–152.

Wilson, T. L., Miller, P. J., 1978: Two-dimensional plume in uniform groundwater flow, *Journal of the Hydraulics Division, Proceedings of the ASCE*, 104, HY 4, 503–514.

Index

adsorption xiv, 112–118, 117T, 130–131T
aquitard 2, 8, 14, 16, 39, 62, 63, 67, 71, 77, 83, 98, 108–112

biodegradation xiv, 112

capillary fringe xiv, 6, 10, 14, 19, 34, 36, 39–41, 55, 65, 67, 68, 77–78, 85, 86, 92, 96–98
concentrations, reasons for low 104
concrete 61, 126

density of pure solvents 130–131T
  of air saturated with solvents 113T, 130–131T
density, effects of, of pure solvents xiv, 23
  of dissolved solvents 7, 13, 100–102
  as gaseous solvents 6, 121–122
diffusion in the gas phase xv, 6, 8, 77–78, 119–123
  in the liquid phase 72
dispersion xiv, 8, 99, 101–103, 113
dissolution xiv, xv, 1, 2, 7, 8, 33, 83, 99-118, 110T
dolomite 66
drinking water standards xiv

fingering 10, 16, 21, 29, 30, 31, 33, 36, 37, 54, 56-58, 62, 65, 67, 69, 70, 71, 93–95
fractured media 2, 61–72, 95–98, 103

glass frits 25T, 24–28

Henry's Gas Law Constants 130–131T
heterogeneities 99
hydraulic roughness, effects of 61–72, 95, 96

infiltration 33, 36, 37, 107
insular 39, 126
irreducible saturation 57, 58

karst 61

limestone 66
lysimeters 28–37

miscibility 1, 37, 57–60
multi-phase flow 2, 57–60, 104

non-wetting phase 50, 51, 55, 57, 59, 87–94, 126

pendular 39, 126
permeability, effects of 2, 6, 7, 8, 10–14, 16, 20, 22–32, 35, 36, 39, 41, 44, 45, 49, 56, 77–80, 93, 107T,112

quartzite 66

recovery 17, 32, 33, 36, 107, 108–112, 110T, 126
residual saturation 2, 7, 8, 16, 39–45, 45T, 53, 54, 57–60, 69, 70, 96, 103–107, 126
retardation 116–118, 117T
retention capacity 32, 33, 34, 39–45, 67, 69, 72, 126

sandstone 66
solubility 130–131T
spills 2, 15
Superfund xiii
surface tension 1, 2

trihalomethanes xxi

vaporization xiv, xv, 1, 2, 6, 8, 18, 70, 77–78, 126
viscosity xiv, 10, 17, 60, 85, 86, 130–131T

wall effects 23, 31, 35
wetting phase 50, 51, 55, 59, 87–94, 126